climate and development

Editorial board

VOLUME 3 ISSUE 1 2011

climate and development

Sustainable adaptation to climate change: Prioritising social equity and environmental integrity

Cover image: Panos/Philippe Lissac

Climate and Development 3(1) January 2011.

First published in paperback 2024

First published by Earthscan in the UK and USA in 2011

For a full list of publications please contact:
Earthscan
4 Park Square, Milton Park, Abingdon, Oxon OX14 4RN
605 Third Avenue, New York, NY 10158

Earthscan is an imprint of the Taylor & Francis Group, an informa business

Publisher's Note
The publisher has gone to great lengths to ensure the quality of this reprint but points out that some imperfections in the original copies may be apparent.

Climate and Development is published by Earthscan and is gratefully supported by the Swedish International Development Cooperation Agency (Sida) and the Stockholm Environment Institute (SEI) in partnership with the Global Change System for Analysis, Research and Training (START). It is editorially impartial and its editors make their own decisions, independent of the journal's supporting organisations.

ISBN: 978-1-84971-413-6 (hbk)
ISBN: 978-1-03-293121-0 (pbk)
ISBN: 978-1-84977-691-2 (ebk)

DOI: 10.4324/9781849776912

Sida
SEI STOCKHOLM ENVIRONMENT INSTITUTE

earthscan from Routledge

climate and development

Aims and scope

Climate and Development is dedicated to the range of issues that arise when climate variability, climate change and climate policy are considered along with development needs, impacts and priorities. It aims to make complex analysis of climate and development issues accessible to a wide audience of researchers, policymakers and practitioners, and to facilitate debate between the diverse constituencies active in these fields throughout the world.

The journal provides a forum to communicate research, review and discussion on the interfaces between climate, development, policy and practice. Every three months it presents conceptual, policy-analytical and empirical studies of the interactions between climate impacts, mitigation, adaptation and development on scales from the local to global. Contributions from and about developing countries are particularly encouraged; however, research on developed countries is welcome provided that the link between climate and development is the central theme.

Climate and Development is of direct and vital relevance to academics, policy analysts, consultants, negotiators, industrial and non-governmental organisations, and to all those working to ensure a better understanding of the links between climate and development.

The journal is the platform of choice for academic debate on issues that link climate and development, and invites contributions on all such issues. These include, but are not limited to:

- The vulnerability of communities to the combined impacts of climate change and non-climatic stresses
- Links between development and building capacity to respond to climate change
- The integration (mainstreaming) of climate policy adaptation and mitigation into sectoral planning and development policy
- Conflicts and synergies between mitigation, energy development and poverty
- The importance of climate and long-term weather forecasting for development
- Responsibilities of developing countries in a post-2012 climate policy regime
- The effects of climate change on meeting the Millennium Development Goals
- The implications for development of the UN Framework Convention on Climate Change and its Kyoto Protocol, as well as all other existing or proposed policy frameworks
- Financing arrangements for adaptation and mitigation in developing countries
- Economic analysis of the effects of climate adaptation and mitigation on developing countries
- Traditional knowledge and local strategies for managing natural resources and coping with climate change
- Forest management and its relationship to mitigation, adaptation and development
- Adaptation, mitigation and the poor

These and other topics are addressed in a number of ways, including:

- Research articles (theoretical developments, concepts and methods, empirical analysis and policy assessments)
- Review articles
- Case studies
- Viewpoints
- Book reviews
- Meeting reports

climate and development

Sustainable adaptation to climate change

SIRI ERIKSEN[1,*] and KATRINA BROWN[2]

[1]Department of International Environment and Development studies – Noragric, Norwegian University of Life Sciences, P.O. Box 5003, NO-1432 Aas, Norway

[2]School of International Development and Tyndall Centre for Climate Change Research, University of East Anglia, Norwich NR4 7TJ, UK

The term 'sustainable adaptation' has emerged with the realization that while adaptation to climate change will be increasingly required over the next decades, we know little about the wider or longer term impacts and implications of adaptation itself. To date there is no certainty that our responses to climate change are sustainable either socially or environmentally, nor how they are likely to contribute to human well-being and poverty alleviation. Previous studies have highlighted how climate change represents both a threat (Yohe et al., 2007) and an opportunity (Commission on Climate Change and Development, 2009) for sustainable development. Only recently have there been attempts to document and compare the experiences of both externally and internally initiated, planned and autonomous, adaptations and their impact in developing countries (McGray et al., 2007; Below et al., 2010; Mearns and Norton, 2010). From these and other studies, there is emerging evidence that many of our responses run counter to principles of sustainable development. This suggests that adaptation policies and interventions that focus on reducing specific climate sensitivities such as predicted changes in precipitation or hydrological regimes, can, even if benefiting some interests, at the same time adversely affect vulnerable groups and create social inequity, as well as unintentionally undermining environmental integrity (Barnett and O'Neill, 2010).

In this special issue of Climate and Development, we argue that radical change is required in order to push for strong sustainability in responses to climate change. Eriksen and O'Brien (2007) highlighted the importance of ensuring that adaptation is socially and environmentally sustainable, contributing to poverty reduction as well as confronting the socio-environmental processes driving vulnerability. To ensure this, and in order to respond to climate change in ways that contribute to sustainable development, there is an urgent need to integrate the relatively well-established understanding of sustainable development with the more recent concerns of vulnerability to climate change. Academic thinking and scientific analysis are now taking up this challenge, and a double session at the Human Security in an Era of Global Change Conference in June 2009 was devoted specifically to presentations and discussions on 'sustainable adaptation'.

This issue synthesizes this discussion, bringing together a collection of articles – originally presented at the 2009 conference – that discuss the challenge of ensuring that we respond to climate change in a way that contributes to sustainable development, in terms of both equity and environmental integrity in the long term. The theme that we wish to highlight is that how we adapt to climate change – in terms of the types of measures, policy frameworks and local household strategies – matters for future development, and particularly for social and environmental sustainability. This has been largely overlooked, and adaptation is normally assumed to be benign for development. The articles explore some of the problems with such an assumption, and what directing adaptation towards principles of sustainability would mean in practice, as well as the conceptual and practical challenges inherent in trying to make adaptation sustainable. The articles develop our understanding of sustainable adaptation by both including conceptual discussion and drawing on empirical research from diverse geographical contexts, including Kenya, Vietnam and the UK. This collection hence addresses a shortcoming in adaptation literature so far; that is, the tendency to consider adaptation in developing and developed countries as very separate issues, when human responses are in fact closely interlinked through their direct and indirect effects on other groups and regions.

*Corresponding author. ***E-mail:*** **siri.eriksen@umb.no**

ISSN: 1756-5529 (print)

The first article (Eriksen et al., 2011) discusses the conceptual underpinnings of the term sustainable adaptation and its practical implications. Initially, sustainability was associated with emission reductions (Markandya and Halsnaes, 2002). With the realization that we need to be critical of *how* we adapt in terms of consequences for poverty and environmental challenges facing the world, sustainability has also been linked to adaptation. Few studies or projects so far have paid explicit attention to what the implications of climate responses will be for social equity and justice and environmental integrity, whether these responses are local and autonomous in nature or planned policy initiatives. Eriksen et al. argue that not every response is necessarily a good one: trade-offs and the potential for negative outcomes over space and time must be recognized. Furthermore, adaptation is a process rather than a list of actions and measures, and consequences of actions and measures must be considered within the much broader social and environmental context. Fundamental societal transformations are required in order to achieve sustainable development pathways and avoid adaptation funding going into efforts that exacerbate vulnerability and contribute to rising emissions. Sustainable adaptation is likely to entail forms of societal organization that are flexible in the face of changing climatic conditions, while at the same time minimizing greenhouse gas emissions. Eriksen et al. outline four principles as a first step in guiding climate responses towards principles of social justice and environmental integrity: first, recognize the context for vulnerability, including multiple stressors; second, acknowledge that differing values and interests affect adaptation outcomes; third, integrate local knowledge into adaptation responses; and fourth, consider potential feedbacks between local and global processes.

Brown's article continues in a more conceptual vein and explores the relationship between sustainable development and adaptation. It questions whether sustainable adaptation may be criticized on the same grounds as sustainable development (Brown, 2011). Are sustainability and adaptation really compatible and can we find policies and measures that will effectively integrate them? It reviews some of the evidence for 'unsustainable adaptation' and discusses how poverty reduction and adaptation goals may be incompatible. These issues pose serious challenges to successful implementation of sustainable adaptation. Furthermore, the article argues that there is a real danger that sustainable adaptation might be co-opted, as sustainable development has been, to promote 'business as usual' rather than to challenge dominant, modernist and technocratic approaches to development. In effect, can the promotion of sustainable adaptation help in shifting attention to the causes of vulnerability and hence the more comprehensive form of climate responses that are now required?

The next three articles present empirical evidence exploring some of the tensions and complexities, competing interests and knowledge, and winners and losers inherent in adaptation and offer perspectives on how we can understand adaptation through a lens of sustainability. As the sustainable development literature has shown, and as these articles demonstrate, ensuring social equity and environmental integrity is not an apolitical or straightforward process, and this has important implications for governance at many scales (Adger and Jordan, 2009). Beckman (2011), writing about Vietnam, considers how different interests may converge and conflict in adaptation to environmental change. The article focuses on policies for adaptation and mitigation that, while increasing resilience at one scale may cause increased vulnerability at other scales. Policies on forest protection and the construction of hydroelectric dams regulate flooding and prevent salt water intrusion into lowland areas and are being justified in part from the perspective of increased awareness of risks associated with climate change. However, the policies also result in severe constraints in access to land for the mountain population, which impacts on their capacity to manage risk and adapt to environmental change. It is argued that these effects are further reinforced by policies of privatization of forest land, reducing the access of the poor to common property resources. Such resources have previously had an important buffer function when coping with crises like serious floods. The resulting difference in adaptive capacity between groups with different access to resources is a critical issue for equity and social sustainability in adaptation. The article represents an important problematization of adaptation, showing that some adaptation may even lead to loss of adaptive capacity in the long term.

Using a case from eastern Kenya, Owuor et al. (2011) demonstrate that sustainable adaptation can be undermined by formal policies and state practice. They investigate how the pastoral and agropastoral populations interact in adapting to climate variability and change, and to drought in particular. They show that interactions in terms of trade and mobility and accessing forest resources are important components of sustainable adaptation in drylands, in terms of reducing the vulnerability of both pastoral and agropastoral groups. However, formal policies and informal governance are impeding these strategies, contributing to conflict between and within agropastoral and pastoral groups, hindering access to natural resources critical for managing drought, hence exacerbating vulnerability and undermining sustainability. The case demonstrates that the extent to which social inequities are created or reinforced through adaptation depends on local politics and governance issues as well as the type of interventions by governmental or non-governmental interventions. Sustainable adaptation for the case of Endau would imply a fundamental change in governance regime from one of imposing punitive measures to stop dynamic interactions to one through which, instead, interactions between the various groups are strengthened.

Adaptation is also a matter of interaction between local strategies to manage change and policies and decision making that may favour particular strategies and developments over others. Exploring some of the local mechanisms, in particular use of indigenous plant resources in dryland areas in Kenya, Gachathi and Eriksen (2011) argue that some of these local practices may represent more environmentally sustainable forms of adaptation to drought and climate change than external or exotic solutions. However, they give very low income and become mechanisms of the poor rather than being able to address existing social inequity. The paper exemplifies the role of indigenous knowledge and resources in striving towards more sustainable forms of adaptation, as well as some of the challenges with achieving this. In particular, it suggests that building on local knowledge, the third principle of sustainable adaptation identified in Eriksen et al. (2010), does not effectively lead to sustainable adaptation unless other principles are also addressed at the same time, such as targeting the vulnerability context and multiple stressors marginalizing dryland populations.

The space and interpretation of sustainability in adaptation and development discourse may be critical for what types of adaptation responses and development paths are supported. Understanding discourse provides important insights into how governance and policies are framed and is as important for developed countries as for developing countries. The final article in the collection, by Opperman (2011), distinguishes the space for sustainability in the discourse of adaptation to climate change in the United Kingdom. She investigates how adaptation is conceptualized in the UK, and that particular framings can limit the possibilities of understanding adaptation more substantively in terms of internal connectivity and complexities. The argument is important for understanding how adaptation can contribute to sustainable development, and that 'sustainable adaptation' in terms of social equity, environmental integrity and conflicts of interest between different actors may in fact not be possible unless existing policy discourses are challenged.

Sustainable adaptation is essentially a global issue since social equity and environmental integrity are important for all populations in order to achieve sustainable responses to climate change. The environmental integrity of adaptation actions in one area also has a bearing on populations elsewhere, for example, through the effect of such actions on emissions and hence on the climate change problem itself. In this rapidly evolving area of both policy and science, the inter- and intro-generational dimensions of climate change responses must be thoroughly and critically interrogated. For sustainable adaptation to avoid becoming another buzzword and for it to meaningfully integrate social, environmental and economic concerns, we should learn from the experiences in different contexts. In documenting the winners and losers, the multi-dimensional trade-offs, and highlighting the competing understandings, meanings and political contestations, this collection of articles is an attempt to initiate learning on sustainable adaptation.

The articles here present a series of important challenges to evolving knowledge, policy and practice on adaptation. Clearly, many current adaptations are far from sustainable. Achieving sustainable outcomes requires that wider social, political and cultural issues are taken on board and that adaptation cannot just be about finding technical solutions or 'fixes' to specific climate impacts. Sustainable adaptation necessitates understanding the role of climate change risks and impacts in the context of other societal and environmental changes. For adaptation to be sustainable it cannot just be about projects, nor can it be about amending current action by adding more tick boxes to a screening process. Sustainable adaptation means we must shift the boundaries of current adaptation and address the underlying causes of vulnerability and poverty in devising responses to climate change.

References

Adger, W. N. and Jordan, A. (eds), 2009. *Governing Sustainability*. Cambridge University Press, Cambridge.

Barnett, J. and O'Neill, S., 2010. Maladaptation. *Global Environmental Change*, 20. 211–213.

Beckman, M., 2011. Converging and conflicting interests in adaptation to environmental change in Central Vietnam. *Climate and Development*, 3(1). 32–41.

Below, T., Artner, A., Siebert, R. and Sieber, S., 2010. *Micro-level Practices to Adapt to Climate Change for African*

Small-Scale Farmers. IFPRI Discussion Paper 00953, Environment and Production Technology Department, IFPRI.

Brown, K., 2011. Sustainable adaptation: an oxymoron? *Climate and Development*, 3(1). 21–31.

Commission on Climate Change and Development, 2009. *Closing the Gaps: Disaster Risk Reduction and Adaptation to Climate Change in Developing Countries*. Ministry of Foreign Affairs, Stockholm, Sweden.

Eriksen, S. and O'Brien, K., 2007. Vulnerability, poverty and the need for sustainable adaptation measures. *Climate Policy*, 7. 337–352.

Eriksen, S., Aldunce, P., Bahinipati, C. S., Martins, R. D'A., Molefe, J. I., Nhemachena, C., O'Brien, K., Olorunfemi, F., Park, J., Sygna, L. and Ulstrud, K., 2010. When not every response to climate change is a good one: identifying principles for sustainable adaptation. *Climate and Development*, 3(1). 7–20.

Gachathi, F. and Eriksen, S., 2011. Gums and resins: the potential for supporting sustainable adaptation in Kenya's drylands. *Climate and Development*, 3(1). 59–70.

Markandya, A. and Halsnaes, K. (eds), 2002. *Climate Change and Sustainable Development: Prospects for Developing Countries*. Earthscan, London.

McGray, H., Hammill, A. and Bradley, R., 2007. *Weathering the Storm: Options for Framing Climate Change and Development*. WRI Report. World Resources Institute, Washington, DC.

Mearns, R. and Norton, A. (eds), 2010. *Social Dimensions of Climate Change: Equity and Vulnerability in a Warming World*. The World Bank, Washington, DC.

Opperman, E., 2011. The discourse of adaptation to climate change and the UK Climate Impact Programme: De-scribing the problematization of adaptation. *Climate and Development*, 3(1). 71–85.

Owour, B., Mauta, W. and Eriksen, S., 2011. Strengthening sustainable adaptation: examining interactions between pastoral and agropastoral groups in dryland Kenya. *Climate and Development*, 3(1). 42–58.

Yohe, G. W., Lasco, R. D., Ahmad, Q. K., Arnell, N. W., Cohen, S. J., Hope, C., Janetos, A. C. and Perez, R. T., 2007. Perspectives on climate change and sustainability. *Climate Change 2007: Impacts, Adaptation and Vulnerability. Contribution of Working Group II to the Fourth Assessment Report of the Intergovernmental Panel on Climate Change*, M. L. Parry, O. F. Canziani, J. P. Palutikof, P. J. van der Linden and C. E. Hanson (eds). Cambridge University Press, Cambridge, UK. 811–841.

review article

climate and development

When not every response to climate change is a good one: Identifying principles for sustainable adaptation

SIRI ERIKSEN[1,*], PAULINA ALDUNCE[2], CHANDRA SEKHAR BAHINIPATI[3], RAFAEL D'ALMEIDA MARTINS[4], JOHN ISAAC MOLEFE[5], CHARLES NHEMACHENA[6], KAREN O'BRIEN[7], FELIX OLORUNFEMI[8], JACOB PARK[9], LINDA SYGNA[7] and KIRSTEN ULSRUD[7]

[1]Department of International Environment and Development Studies, Noragric, Norwegian University of Life Sciences, P.O. Box 5003, No-1432, Aas, Norway
[2]Department of Environmental Sciences and Renewable Natural Resources, University of Chile, Sta. Rosa 11.315, La Pintana, Santiago, Chile; Department of Resource Management and Geography, Melbourne School of Land and Environment, University of Melbourne, Melbourne, Victoria 3010, Australia
[3]Madras Institute of Development Studies (MIDS), 79, Second Main Road, Gandhi Nagar, Adyar, Chennai 600 020, India
[4]Center for Environmental Studies (NEPAM), University of Campinas (UNICAMP), Cid. Univ. Zeferino Vaz, Campinas, SP 13083-867, Brazil
[5]Department of Environmental Science, University of Botswana, Private Bag 00704, Gaborone, Botswana
[6]Council for Scientific & Industrial Research, Meiring Naude Road, Pretoria 0001, South Africa
[7]Department of Sociology and Human Geography, University of Oslo, P.O. Box 1096, Blindern, Oslo 0317, Norway
[8]Nigerian Institute of Social and Economic Research, PMB 05, Ojoo, Ibadan, Nigeria
[9]Green Mountain College, One Brennan Circle, Poultney, VT 05764, USA

Climate adaptation has become a pressing issue. Yet little attention has been paid to the consequences of adaptation policies and practices for sustainability. Recognition that not every adaptation to climate change is a good one has drawn attention to the need for sustainable adaptation strategies and measures that contribute to social justice and environmental integrity. This article presents four normative principles to guide responses to climate change and illustrates the significance of the 'sustainable adaptation' concept through case studies from diverse contexts. The principles are: first, recognize the context for vulnerability, including multiple stressors; second, acknowledge that differing values and interests affect adaptation outcomes; third, integrate local knowledge into adaptation responses; and fourth, consider potential feedbacks between local and global processes. We argue that fundamental societal transformations are required in order to achieve sustainable development pathways and avoid adaptation funding going into efforts that exacerbate vulnerability and contribute to rising emissions. Despite numerous challenges involved in achieving such change, we suggest that sustainable adaptation practices have the potential to address some of the shortcomings of conventional social and economic development pathways.

Keywords: adaptation; climate change; environmental change; sustainable development; transformation; vulnerability

1. Introduction

Climate adaptation has become a more visible and pressing issue in recent years. In part this can be attributed to the recognition that the climate system will undergo changes in the coming century regardless of reductions in greenhouse gas emissions, mainly due to thermal inertia of oceans and the long atmospheric lifetime of carbon dioxide and other greenhouse gases (Matthews and Caldeira, 2008). However, it has also been reluctantly acknowledged

*Corresponding author. *E-mail:* **siri.eriksen@umb.no**

ISSN: 1756-5529 (print)

that emission reductions are unlikely to decrease at the rate and magnitude necessary to prevent climate change that is dangerous to many (Parry et al., 2009; Schellnhuber, 2009). Adaptation is thus increasingly considered as essential to reducing vulnerability to dangerous climate change.

Yet, although adaptation can potentially reduce the negative impacts of climate change, little attention has been paid to the consequences of adaptation policies and practices for sustainability. In some cases, what seems to be a successful adaptation strategy to climate change may in fact undermine the social, economic and environmental objectives associated with sustainable development. Strategies or policies that make sense from one perspective, or for one group, may at the same time reduce the livelihood viability or resource access of other groups. Likewise, an eagerness to reduce climate risk through specific technologies or infrastructural changes may sometimes lead to the neglect of other environmental concerns, such as biodiversity (Næss et al., 2005; Eriksen and O'Brien, 2007; Eriksen and Lind, 2009). Hence, adaptation can have unintended negative effects both on people and on the environment.

A recognition that not every adaptation to climate change is a good one has drawn attention to the need for sustainable adaptation strategies and measures, and for qualifying what types of adaptation are desirable or not (Eriksen and O'Brien, 2007). There is also an increasing recognition of the potential of climate adaptation to address some of the mistakes and shortcomings of conventional social and economic development pathways that have contributed to social inequity, poverty and environmental problems (Ulsrud et al., 2008). It is particularly important to identify the synergies between adaptation and sustainable development because urgent and overwhelming poverty problems in the world are far from satisfactorily addressed, and environmental problems other than climate change also threaten people's livelihoods and quality of life. Indeed, most individuals and communities are adapting to multiple stressors, in addition to climate variability, extremes and the risk of disaster (Eakin, 2006; Reid and Vogel, 2006; Schipper and Pelling, 2006; Ziervogel et al., 2006; O'Brien et al., 2008).

Developed countries are committed to the goal of jointly mobilizing USD30 billion for the period 2010–2012 (and an additional USD100 billion a year by 2020) to address the climate-related challenges of developing countries, and much of this will go to adaptation (ENB, 2009). The increase in attention to and resources for adaptation suggests that it is critical to 'get adaptation right' in order to solve, rather than exacerbate, problems. Consequently, it is important to understand what it means to sustainably adapt to climate change, or what is referred to in this article as 'sustainable adaptation'. Sustainable adaptation is defined here as adaptation that contributes to socially and environmentally sustainable development pathways, including both social justice and environmental integrity.

This article presents and discusses the concept of sustainable adaptation to climate change and identifies four normative principles to guide responses to climate change. We illustrate the principles of sustainable adaptation and their significance through case studies from diverse contexts. In the conclusions, we discuss the possibilities and limitations for achieving sustainable adaptation in practice. We suggest that despite numerous challenges, attention to principles for sustainable adaptation can contribute to socially and environmentally sustainable responses to climate change.

2. Climate change adaptation and sustainable development

Adaptation to climate change has been described from a wide range of perspectives, and many adjectives have been used to modify the term (autonomous, involuntary, planned, passive, reactive or anticipatory, etc.). In terms of climate change, adaptation has been defined as the process or adjustments through which people reduce the adverse effects of climate on their health and well-being, and take advantage of the opportunities that their climatic environment provides. Other definitions have argued more forcefully that adaptation includes the reduction of vulnerability (Smit et al., 2000; Debels et al., 2009). Leary (1999) and Burton et al. (2002) referred to climate adaptation as a wide range of behavioural adjustments that households and institutions make (including practices, processes, legislation, regulations and incentives) to mandate or facilitate changes in socio-economic systems, aimed at reducing vulnerability to climatic variability and change. Nelson et al. (2007) defined adaptation as the decision-making process and the set of actions undertaken to maintain the capacity to deal with current or future predicted change. These definitions are summarized in the Intergovernmental Panel on Climate Change (IPCC) definition of adaptation: the adjustment to practices, processes and systems in order to ameliorate negative effects and take advantage of opportunities associated with climate change (IPCC, 2007).

Debates on climate change adaptation have taken place largely outside of the broader discourse on sustainable development (Bizikova et al., 2010). Although sustainable development has been included as a theme in many of the assessments by the IPCC (Munasinghe and Swart, 2000; Yohe et al., 2007), little attention has been paid to the identifying principles that create synergies between adaptation and sustainable development. Cohen et al. (1998) pointed out that although climate change is one of the most important symptoms of an unsustainable economic system, the climate change and sustainable development fields have been separated by differences in discourse. For example, climate change has been largely constructed as an environmental problem that can be solved by reducing greenhouse gas emissions, with little attention to its social, cultural, political and ethical dimensions (O'Brien et al., 2010). This effectively bypasses the complex, context-specific and multidimensional challenges of sustainable development. The concept of sustainable development initially focused on the close connection between environmental problems, poverty, inequity and basic human needs. However, the concept of sustainability has been criticized as a vague policy term rather than an academic concept subject to rigorous analysis. It has been accused of being malleable to suit any interest, or a 'rhetorical cover for business-as-usual politics' (Cohen et al., 1998, p. 353), distracting attention from any fundamental changes in systems. There have, however, been many calls for 'strong sustainability', which involves changing current modes of development, questioning calls for continued economic growth and appealing for a less managerial approach to human–environment relations (Adams, 2009).

Cohen et al. (1998) argued that it is precisely in forging the links between climate change and sustainable development, in terms of focusing rigorous analysis and policy efforts on the political, social and ethical dimensions, that action in both areas can be achieved. According to Robinson and Herbert (2001), climate change can be made more relevant to policy by contextualizing it within a sustainable development framework. They argue that mitigation and adaptation can contribute to a range of sustainability goals, at the same time that sustainable development policies can contribute to emission reductions. As with debates about sustainable development, the climate change problem raises questions about the underlying development pathways causing both environmental problems and poverty (Adams, 2009). The issues of climate change and sustainable development thus converge in the call for fundamental changes to development pathways. A critical point is the recognition of alternative development paths, and 'how much choice we have about what kind of world we will end up in' (Robinson and Herbert 2001, p. 146).

3. Key principles for sustainable adaptation

An underlying premise for the concept of sustainable adaptation is that many responses to climate change will create social and environmental externalities, including trade-offs and negative consequences. Sustainable adaptation thus considers the wider effects of adaptive responses on other groups, places and socio-ecological systems, both in the present and in the future. Sustainable adaptation can be distinguished from adaptation in general in that it qualifies actions in terms of their effects on social justice and environmental integrity; that is, adaptation is sustainable only if it contributes (and at the very least does not seriously erode) these two features. This qualifying of adaptation is a response to concerns that adaptation has often been operationalized in practice through changes in technology, institutions and managerial systems (Klein et al., 2007), rather than challenging current development paths, including the social, economic and political structures that underlie many contemporary problems.

Sustainable adaptation can be considered necessary in response to three problems highlighted in the vulnerability literature. First, climate change is a global problem that affects both current and future generations, and responses must be sensitive to both spatial and temporal consequences. Adaptations taken to benefit one sector or group may undermine the security and well-being of others, such as by influencing resource access and the integrity of ecosystems that many people depend upon for their livelihoods (Eriksen et al., 2005). Second, widespread poverty makes many individuals, households, communities and states vulnerable to even small shocks and stressors. The tendency of poor people to be highly vulnerable to climate change is often used as a justification for implementing adaptation; however, whether or not the proposed adaptation measures will actually assist poor groups is seldom assessed. Since not any and every adaptation intervention reduces poverty and inequality (and some poverty reduction measures may aggravate vulnerability), sustainable adaptation measures need to specifically target links between vulnerability and poverty (Eriksen and O'Brien, 2007; Eriksen et al., 2007). Third, the need to drastically reduce global greenhouse gas emissions and facilitate a rapid transition to low-emission economies suggests that adaptation measures should emphasize low-emission solutions. Responses to climate change can thus be seen as a means for promoting alternative development pathways, such as transitions to low-carbon economies, organic agriculture and horticulture, agroforestry, ecological sanitation, water harvesting, water purification by the use of solar energy, alternative modes of transport, decentralized renewable energy supply, recycling or participatory plant breeding (Ulsrud et al., 2008; Winkler and Marquand, 2009).

Sustainable adaptation differs from a reformist view of sustainable development, and from an interpretation of adaptation as a mere adjustment of current practices and development paths. For example, development paths that contribute to inequity and poverty, or are based on fossil fuel-intensive consumption patterns, are inevitably called into question by the concept of sustainable adaptation. The types of responses that contribute to social equity and environmental integrity will depend on the context, and therefore vary between people and places, and over time. Hence, 'sustainable adaptation' does not suggest that a specific technology or practice can be identified that will be viable in all places or at all times. Instead, practices need to change as the context changes, forming part of the new and dynamic development paths required to reduce both vulnerability and greenhouse gas emissions.

The question then arises as to what characteristics or conditions should be looked for when assessing adaptation responses? How can the concept of sustainable adaptation be realized? Four main principles are presented here, and elaborated on through case studies that illustrate how adaptation can be formulated in different contexts. The challenges in using such an approach are also discussed. As with all responses to climate change, it is important to consider the vested interests, the mismatches between the scales of action and issues of power relations, the prioritization of certain types of knowledge and the lack of systems perspective in the process of decision-making. These factors are, however, likely to become more visible if the principles are included in adaptation planning.

3.1. Key principle 1: recognize the context for vulnerability, including multiple stressors

Individuals, groups and regions are experiencing many types of stressors, besides environmental change, that together create a context for vulnerability (Eakin, 2006; Ziervogel et al., 2006; Leichenko and O'Brien, 2008; Eriksen and Lind, 2009; Tschakert and Dietrich, 2010). Recognizing the role of multiple stressors in influencing this context for vulnerability involves acknowledging that despite good intentions, some adaptations may not improve social equity and environmental integrity. The underlying social, economic, institutional and cultural conditions that contribute to a wider context for vulnerability thus need to be understood, in order to identify direct and indirect consequences of adaptation efforts, and to be sensitive to the spatial and temporal effects of such efforts. In terms of social and environmental consequences, sustainable adaptation thus places a greater emphasis on how the structural and contextual factors that create vulnerability, such as chronic poverty and unequal terms of trade, influence the outcomes of adaptation measures. This first principle of sustainable adaptation thus holds that responses should be sensitive to the wider context in which climate change is experienced.

3.1.1. Case study: addressing the vulnerability context of poor communities affected by floods and rainstorms in the city of Ilorin, Nigeria

The importance of this principle is illustrated by the case of poor, urban and semi-urban areas of Ilorin, the capital city of Kwara State in Nigeria. There are multiple stressors that generate vulnerability in these areas, and unless socio-economic dimensions are tackled in combination with infrastructure, climate-related extreme events, such as heavy rainstorms and flooding, will continue to have effects on livelihoods and long-term vulnerability. A large proportion of inhabitants in the case study area are older people (40% are above 50 years of age), levels of education are low and very few are engaged in the formal sector (9%), most working as artisans, farmers and traders. Household sizes are large: close to 80% of the households have more than four people. At the same time, houses are old (more than half are older than 30 years) and many are constructed in materials that do not withstand rainstorms and flooding. Poor waste collection leads to blocked drainage systems. In some parts the situation is made even more precarious due to sparse vegetation, meaning that any heavy rainfall results in flooding (Ijaiya and Umar, 2004). Hence, key conditions generating vulnerability include poverty, overcrowding and social inequity.

A number of socio-environmental changes create the conditions described above. These include the marginalization of urban dwellers in terms of infrastructure, services and income opportunities; rapid urbanization; physical development on environmentally sensitive lands such as wetlands, slopes and floodplains that exacerbates environmental degradation; and flooding risks (Olorunfemi and Raheem, 2007; Olorunfemi, 2008; Mehrotra et al., 2009; Gbadegesin et al., 2010). Extensive damage to properties and livelihoods contribute to the endemic poverty in most parts of Kwara State. For instance, increasingly frequent and severe floods have damaged electricity facilities in some areas for months, disrupted trading, and washed away crops in suburban areas. Traders, artisans and women farmers are among the most vulnerable groups.

In order to develop measures that contribute to sustainable adaptation, it is necessary to address the structural and contextual factors that create vulnerability, such as those described above. Measures also need to include an understanding of how livelihood dynamics form part of the vulnerability context. For example, support from friends and relatives and personal savings explain how a large proportion of disaster victims cope with its immediate impacts. Sustainable adaptation measures must be sensitive to the need to sustain such support networks. At the same time, however, measures would also need to address the vulnerability context in the long term by complementing household mechanisms and addressing some of the structural processes. This could be achieved, for example, by facilitating livelihood diversification and formal support systems that could relieve the stress on social networks in times of disasters. This first principle of sustainable adaptation involves broadening responses to recognize, and where possible address directly, the context in which climate change is experienced. This context includes stressors such as the marginalization of urban dwellers in terms of infrastructure, services and income opportunities, as well as physical developments that threaten environmental integrity and exacerbate flood risk.

3.2. Key principle 2: acknowledge that different values and interests affect adaptation outcomes

Values and interests play an important yet seldom discussed role in climate change responses, and they influence the adaptation strategies that are prioritized by different groups (O'Brien, 2009). Recognizing potential value conflicts can help to identify how adaptation responses taken by one group may affect the vulnerability context of other groups. Strong vested interests within particular adaptation strategies may act as a barrier to sustainable types of adaptation. For example, the adaptive responses that distribute risk across market and subsistence production in Ghana may in fact prioritize the maintenance of the status quo for men, at the cost of women's self-determination (Carr, 2008). Sustainable adaptation may thus involve a more transparent political process that creates enabling conditions and access to information that supports decision-making for adaptation. For example, linking democratization and empowerment efforts with those of adaptation can potentially address differing and often conflicting adaptation interests (Eriksen and Lind, 2009). The second principle involves recognizing differential interests and potential value conflicts, and identifying how these may influence outcomes, particularly for the most vulnerable.

3.2.1. Case study: including the adaptation interests of vulnerable groups in local government policy in Durban, South Africa

The case of Durban, exposed to both flooding and coastal erosion, illustrates how important it is to develop institutions (and how these institutions conceive climate change) that focus on social equity and vulnerability in order to achieve sustainable adaptation. In particular, prioritizing the needs of vulnerable groups in both development and climate policy processes is critical. Before the democratic transition in 1994, environmental concern at the local level was low in South Africa. The process of democratization resulted in a development agenda that focused on the need to address the social inequity created by the Apartheid regime, but with little connection to climate change (Roberts, 2008; Carmin et al., 2009). In the beginning, any climate change action was also largely disconnected from concerns about adaptation and vulnerability; for example, the Cities for Climate Protection campaign initiated in 2000 largely focused on developing mitigation-related policies (Roberts, 2008). Although important as a first step, the campaign failed to generate an institutional framework, knowledge about climate change and adaptation, or interest among government agencies or the population at large (Carmin et al., 2009).

The situation improved when programmes started to focus more specifically on vulnerability and climate protection, such as through convening a vulnerability assessment. This assessment served as an opportunity to engage different municipal stakeholders in climate change discussions (Carmin et al., 2009), leading to recognition of the city's vulnerability and of existing initiatives through which adaptation could be facilitated. A second phase focused on key municipal sectors such as urban infrastructure, human health and disaster risk reduction (Roberts, 2008).

The case indicates that it is important not only to mainstream climate change responses into local government policies but also to consider it under a framework of social inclusion, justice and sustainable development. Not only could the interests of vulnerable groups be heard; by including vulnerable groups in the science–policy interface understanding of the implications of climate change in the local context was enhanced, generating local interest and policy action (Vogel et al., 2007). The case also exemplifies the importance of having local champions within government structures that can spearhead such engagement, an observation previously made in other contexts such as Norway, Sweden and the USA (Næss et al., 2005; Lowe et al., 2009; Sanchez-Rodriguez, 2009; Storbjørk et al., 2009). Such a dependence on individuals within government structures can nevertheless be a barrier to the social inclusion of vulnerable groups, since how (and if) processes are designed and which interests are heard are related to the particular knowledge, connections and orientation of an individual rather than institutionalized and democratic adaptation policy processes. The second principle suggests the need to ensure that representation of groups that are vulnerable to climate variability and change is institutionalized in formal government or development processes. It also requires that such processes analyse and recognize different interests and potential value conflicts up front, and identify how these may influence outcomes.

3.3. Key principle 3: integrate local knowledge into adaptation responses

Different groups and actors produce different knowledge on adaptation, and which source of knowledge is recognized and used in decision-making is crucial in determining which interests or development paths are prioritized. Different approaches to adaptation often reflect varying approaches to knowledge and understandings of the local context, resulting in different diagnoses of both problems and solutions. Integrating local knowledge based on the experience of living in a risky place and of observing the natural environment is essential for sustainable adaptation to climate change (Olsson and Folke, 2001; Berkes, 2007). Community-based adaptation initiatives are increasing in response to the top-down, technical approaches promoted by the scientific discourse on climate change (Huq and Reid, 2007). In the dominant scientific discourse, practices of the poor have often been blamed for environmental degradation, and resource control has consequently been transferred from local populations to central governments or to private actors (Benjaminsen et al., 2006). The third principle of sustainable adaptation recognizes that successful responses involve integrating local knowledge with other sources of knowledge about climate change.

3.3.1. Case study: building on local knowledge and capacity in risk reduction in Concepción, Chile

The importance of existing local knowledge and capacity is particularly well illustrated by the case of Concepción, Chile.[1] Over time, vulnerable people have developed responses to disasters based on their knowledge and understanding of the conditions and environment where they live. The community of Agüita de la Perdiz consists of mainly informal and illegal settlements, built on landslide-prone areas on the 'Caracol hill', downtown of the second largest city in Chile, Concepción (Mardones and Vidal, 2001; Hauser, 2005).

Climate-related hazards, such as rainfall or cyclones, are expected to increase in frequency and magnitude because of climate change. However, there remains substantial uncertainty in the rate and behaviour of these changes (Christensen et al., 2007). Hence, timely and local adaptation to 'new unknown severity and frequency of hazards' under a changing climate becomes imperative (Debels et al., 2009). In 2005, the community living in this area faced the most severe event in 142 years when 162.2 mm precipitation fell in 24 h. The material damages were massive, with 100 out of 282 houses partially or completely destroyed (DMC, 2005; ONEMI, 2005). What was remarkable for a disaster of this magnitude was that there were no deaths reported, and only a few injuries. In-depth interviews with people affected by the flood revealed that a crucial aspect that helped to protect what is most important – their lives – was the knowledge people had of their environment and vulnerability (Aldunce et al., forthcoming).

Recognizing and acting on an unusual level of rainfall, the community made use of both past experience and knowledge about which areas would be most exposed and which people would be hardest hit. Rather than waiting for external warning and help, people organized a refugee camp, evacuated vulnerable community members and took turns to protect houses against robbery (Aldunce et al., forthcoming). Faced with recurrent extreme events, the Agüita de la Perdiz community has shown itself capable of generating social learning, and the population has a high level of risk awareness and knowledge about the physical environment and potential vulnerability. This in turn has resulted in proactive behaviour in terms of well-organized community participation and leadership in disaster response, and improved capacity to adapt to climate extremes. The high degree of social learning enabled people of Agüita de la Perdiz to assist neighbouring communities in their response and recovery, both during the 2005 deluge and in other disasters. The key role of autonomous adaptation and local knowledge in adapting to climate variability and change has been frequently illustrated in rural contexts (Eriksen et al., 2005; Eakin, 2006; Reid and Vogel, 2006; Ziervogel et al., 2006). The case of Agüita de la Perdiz shows the more universal relevance of the third principle of sustainable adaptation for both urban and rural contexts; that is, the importance of generating local knowledge and integrating it with other sources of knowledge in order to develop successful responses to climate change and empower local decision-making. Local knowledge in disaster risk management is critical for reducing vulnerability among the poorest, and can be combined with policy efforts to address social equity and vulnerability. Any policy intervention to strengthen adaptation and reduce risk would need to recognize community participation in disaster prevention and response and strategies for living with environmental variability (Wisner et al., 2004; Eriksen et al., 2005; Pelling and High, 2005; van Aalst et al., 2008).

3.4. Key principle 4: consider potential feedbacks between local and global processes

Adaptation responses may directly affect the vulnerability of local populations, but every response can also influence – or be influenced by – larger-scale processes. As Adger et al. (2009) pointed out, vulnerability is nested and tele-connected through environmental change feedbacks, economic linkages and global flows of resources, people and information. The possibility that feedbacks and linkages can influence both social justice and environmental integrity over both space and time raises questions about the sustainability of many adaptation responses. For example, adaptations often have significant implications for greenhouse gas emissions, water quality and access, and biodiversity. Likewise, adaptations can influence migration, trade patterns and urbanization processes. Mitigation of climate change is particularly important, as continued global warming can overwhelm local adaptive capacity. The fourth principle of sustainable adaptation hence focuses on the need for responses to recognize the interactions between local and global processes, which can create both positive and negative feedbacks.

3.4.1. Case study: linking adaptation with mitigation and transformations towards a resilient society in Norway

The importance of embedding local actions and adaptation in an understanding of climate change as a global concern is illustrated in the case of snow-dependent leisure activities in Oslo, Norway. For local adaptation efforts to be considered sustainable there is the need to consider the global effects of these efforts. For example, using low- rather than high-energy adaptation options would limit greenhouse gas emissions that contribute to global warming and increased risk elsewhere.

Winter sports and leisure activities such as skiing and skating are ingrained in the Norwegian national identity. A warming climate has led to deteriorating snow and ice conditions, especially since the 1970s. In the Oslo region, inhabited by a fifth of the country's population[2] and where an estimated 80% use the forests for recreation (Berg, 2004; Vaage, 2004), the number of days with skiing conditions are projected to decline by 40% from the 1981–1999 period to 2050 (Iversen et al., 2005). A transformation of recreational activities and ways of defining national identity may be required in the long term.

However, current adaptations in the face of warming conditions appear to focus on preserving existing activities through 'controlling' local environmental conditions in the short term in the face of changing weather conditions, often in ways that involve increased energy use. For example, the municipal authorities and sports clubs now produce large quantities of artificial snow and ice. In western Oslo, for example, there are now plans to construct the country's biggest artificial ice rink to enable people to skate despite warming winter conditions. Those opposed are concerned about local increases in traffic, noise and light pollution. Completely absent from the debate, however, are concerns about the global climate with respect to the increased emissions that result from the energy used in producing artificial ice. The main climatic consideration in the debate was the potential for local cooling due to the artificial ice.[3] At this instance, local adaptation is clearly not placed in a global context. There is little awareness on how the effects of local adaptation responses, through local and global linkages and feedback processes, in turn affect global warming.

Even if energy consumption is increasingly considered in the production of artificial snow and ice, there are nevertheless limits to such forms of adaptation. The production of artificial snow and ice can only support skating and skiing in isolated areas, while the loss of natural winter conditions and associated recreational activities could damage cultural and emotional attachment to the winter landscape, and potentially lead to a loss of values around national identity.

Sustainable adaptation in the case of Norway would involve both drastic cuts in GHG emissions to reduce future deterioration of snow conditions as well as transformation towards new types of recreation and cultural identities. In the current framing of the climate change problem, however, local weather and responses are treated as isolated from global changes. Such an approach may reinforce a dominant complacency regarding Norway's ability to adapt its way out of climate change (O'Brien et al., 2006) and stifle public and policy engagement for addressing climate change. The fourth principle of sustainable adaptation – recognizing the interactions between local and global processes – involves broadening responses from narrow short-term goals to instead helping to transform society through enhanced resilience and flexibility in the face of uncertainty, accommodating diverse needs (beyond skiing), and recognition of both positive and negative feedbacks from local measures.

4. Conclusions: practical and conceptual lessons regarding sustainable adaptation

Sustainable adaptation can be defined as a set of actions that contribute to socially and environmentally sustainable development pathways, including social justice and environmental integrity. However, just as adaptation provides an opportunity to transform society towards sustainability goals, adaptation actions can also exacerbate greenhouse gas emissions, vulnerability to climate change and a number of development problems. In this article, we have outlined four principles that can guide adaptation responses in a manner that supports sustainability. Sustainable adaptation should (1) recognize the context of vulnerability, including multiple stressors, (2) acknowledge that different values and interests affect adaptation outcomes, (3) integrate local knowledge into adaptation responses and (4) consider potential feedbacks between local and global processes. An underlying premise for the four principles is that adaptation is not neutral, and not all adaptation will 'do good'; there will be trade-offs, feedbacks and negative consequences. Assessing and understanding these dimensions and moving towards sustainable development pathways requires a renewed focus on the consequences of adaptation actions, whether these actions are policy driven or autonomous, or involve social development, altered technology and practice, economic or institutional measures, legislation or infrastructure, or changes to political, structural or social relations.

The four case studies presented above illustrate different aspects of these principles. However, no single case illustrates a perfect or comprehensive example of sustainable adaptation. It is important to acknowledge that even if applied, the four principles alone do not guarantee sustainable adaptation. This article represents a first step in defining sustainable adaptation, and there is clearly a need for continued reflexivity, and what Tschakert and Dietrich (2010) refer to as 'anticipatory learning'. Furthermore, many gaps still exist between research and practice. How, then, can these principles be used to implement sustainable adaptation in practice? While answering this question is beyond the scope of this article, a few reflections are offered below.

Sustainable adaptation is likely to entail societal organization that is flexible in the face of changing climatic conditions (rather than 'controlling' specific environmental conditions), while at the same time minimizing greenhouse gas emissions. It is important that adaptation actions do not lock people into high-emission and soon-obsolete technologies or practices, nor reinforce dependency relations. Instead, actions need to contribute to a cleaner, greener and more equitable society. Navigating the global long-term consequences of adaptation actions is complex. In the case of biofuel production, sustainability would entail promoting energy access and livelihood options by the poor in ways that enhance adaptive capacity, while avoiding production patterns that entrench dependency or create vulnerability, environment and land loss problems (African Biodiversity Network, 2008).

Sustainable adaptation also calls for a strengthening of social resilience. The case of Concepción underscores the importance of social capital and community empowerment as part of sustainable adaptation, through strong citizen participation, local identity and local organization. Social capital is made up of different norms and networks that enable people to act collectively (Woolcock and Narayan, 2000; Adger, 2003) and enable the knowledge sharing, spreading of risk and claims for reciprocity in times of crisis. Such networks are scale dependent and are associated with a flexible and adaptive society (Adger, 2003). A central challenge reflected in the cases, however, is that strengthening local capacity alone does not effectively reduce vulnerability. Increased sustainability can only be achieved if local capacity is combined with measures aimed at including socially marginalized groups, making the voices of vulnerable groups heard in decision-making processes that affect their adaptation interests and making these interests count in the face of pressures from economic development, such as physical development of lands that currently increase climate risk and reduce land rights of the poor.

The road to sustainable adaptation starts with the understanding that adaptation is a 'process' rather than a list of actions and measures that address specific climate change impacts. Sustainable adaptation requires going beyond one-time climate proofing measures, and questioning the assumption that every adaptation to climate change will be beneficial. The consequences of actions and measures must be considered within the much broader social and environmental context; trade-offs and the potential for negative outcomes over space and time must be recognized. The normative principles of sustainable adaptation can be considered a first step in guiding responses towards social justice and environmental integrity.

Acknowledgements

This article is the result of discussions by a group of scientists from Asia, Africa, Latin America, North America and Europe at several meetings, including an International Human Dimension Workshop (IHDW) on Sustainable Adaptation held in New Delhi in October 2008, the IHDP Open Meeting in Bonn 2009, and the GECHS conference in Oslo, June 2009. We are grateful for support for these events from IHDP and the Research Council of Norway. Rafael D'Almeida Martins acknowledges the financial support of the Brazilian Federal Agency for Support and Evaluation of Graduate Education (CAPES) and the State of Sãu Paulo Research Foundation (FAPESP) as well as the hospitality granted by the Department of Environmental Policy Analysis, Institute for Environmental Studies (IVM), Vrije Universiteit Amsterdam.

Notes

1. This case study draws upon on the research carried out by Aldunce, P. and Levín, V. between 2005 and 2007 (Aldunce et al., forthcoming).
2. Statistics Norway: www.ssb.no/utlstat/tab-2009-03-12-05.html, www.ssb.no/utlstat/tab-2009-0.
3. Newspaper articles: www.oslogk.no/Dokumenter/Bogstad%20vinterparadis%20faktaark.pdf; www.akersposten.no/apps/pbcs.dll/article?AID=/20081002/NYHETER/899389345/1052.

References

Adams, W. M., 2009. *Green Development. Environment and Sustainability in a Developing World.* Routledge, London.

Adger, W. N., 2003. Social aspects of adaptive capacity, climate change, adaptive capacity and development. *Climate Change, Adaptive Capacity and Development*, J. B. Smith, R. J. T. Klein and S. Huq (eds). Imperial College Press, London, UK.

Adger, W. N., Eakin, H. and Winkels, A., 2009. Nested and teleconnected vulnerabilities to environmental change. *Frontiers in Ecology and the Environment*, 7(3). 150–157.

African Biodiversity Network, 2008. Letter to Members of the Industry, Research and Energy Committee of the European Parliament: Renewable Energy Directive Must Not Promote Unsustainable Biofuels. *African Biodiversity Network*, 28 August.

Aldunce, P., Levín, V. and León, A., forthcoming. Community participation: a bridge for disaster risk management and adaptation to climate change. *A Changing Environment for Human Security: New Agendas for Research, Policy and Action*, K. O'Brien and L. Sygna and J. Wolf (eds). Earthscan, London.

Benjaminsen, T. A., Rohde, R., Sjaastad, E., Wisborg, P. and Lebert, T., 2006. Land reform, range ecology, and carrying capacities in Namaqualand, South Africa. *Annals of the Association of American Geographers*, 96(3). 524–540. doi: 10.1111/j.1467-8306.2006.00704.x.

Berg, N., 2004. *Holdnings og brukerundersøkelse i Oslo kommune for Friluftsetaten. Utarbeidet for Oslo Kommune, Friluftsetaten (Survey of Attitudes and Users in Oslo Municipality).* Oslo Municipality, Norway.

Berkes, F., 2007. Understanding uncertainty and reducing vulnerability: lessons from resilience thinking. *Natural Hazards*, 41. 283–295.

Bizikova, L., Burch, S., Cohen, S. and Robinson, J., 2010. Linking sustainable development with climate change adaptation and mitigation. *Climate Change, Ethics and Human Security*, K. L. O'Brien et al. (eds). Cambridge University Press, Cambridge. 157–179.

Burton, I., Huq, S., Lim, B., Pilifosova, O. and Schipper, E. L., 2002. From impacts assessment to adaptation priorities: the shaping of adaptation policy. *Climate Policy*, 2(2–3). 145–159. doi: 10.1016/S1469-3062(02)00038-4.

Carmin, J. A., Roberts, D. and Anguelovski, I., 2009. Planning climate resilient cities: early lessons from early adapters. *Proceedings of the World Bank Fifth Urban Research Symposium on Cities and Climate*

Change: Responding to an Urgent Agenda, Marseille, France, 5–8 June.

Carr, E., 2008. Between structure and agency: livelihoods and adaptation in Ghana's central region. *Global Environmental Change*, 18. 689–699.

Christensen, J. H., Hewitson, B., Busuioc, A., Chen, A., Gao, X., Held, I., Jones, R., Kolli, R. K., Kwon, W.-T., Laprise, R., Magaña Rueda, V., Mearns, L., Menéndez, C. G., Räisänen, J., Rinke, A., Sarr, A. and Whetton, P., 2007. Regional climate projections. *Climate Change 2007: The Physical Science Basis. Contribution of Working Group I to the Fourth Assessment Report of the Intergovernmental Panel on Climate Change*, S. Solomon, D. Qin, M. Manning, Z. Chen, M. Marquis, K. B. Averyt, M. Tignor and H. L. Miller (eds). Cambridge University Press, Cambridge, UK and New York, NY, USA, 847–940.

Cohen, S., Demeritt, D., Robinson, J. and Rothman, D., 1998. Climate change and sustainable development: towards dialogue. *Global Environment Change*, 8(4). 341–371. doi: 10.1016/S0959-3780(98)00017-X.

Debels, P., Szlafsztein, C., Aldunce, P., Neri, C., Carvajal, Y., Quintero-Angel, M., Celis, A., Bezanilla, A. and Martinez, D., 2009. IUPA: a tool for the evaluation of the general usefulness of practices for adaptation to climate change and variability. *Natural Hazards*, 50. 211–233. doi: 10.1007/s11069-008-9333-4.

DMC (Dirección Meteorológica de Chile), 2005. *Boletín climatológico*, Vol. XXI(6), Junio de 2005. Gastón Torres (ed.). Santiago, Chile. 19 p.

Eakin, H., 2006. *Weathering Risk in Rural Mexico. Climatic, Institutional, and Economic Changes*. The University of Arizona Press, Tucson.

ENB, 2009. A brief analysis of the climate change conference. *Earth Negotiations Bulletin*, 12(459). 27–29. www.iisd.ca/climate/cop15.

Eriksen, S., Brown, K. and Kelly, P. M., 2005. The dynamics of vulnerability: locating coping strategies in Kenya and Tanzania. *Geography Journal*, 171(4). 287–305. doi: 10.1111/j.1475-4959.2005.00174.x.

Eriksen, S., Klein, R.J.T., Ulsrud, K., Næss, L. O. and O'Brien, K., 2007. *Climate Change Adaptation and Poverty Reduction: Key Interactions and Critical Measures*. Report prepared for the Norwegian Agency for Development Cooperation (Norad). GECHS Report 2007:1, University of Oslo. 42 p.

Eriksen, S. and Lind, J., 2009. Adaptation as a political process: Adjusting to drought and conflict in Kenya's drylands. *Environmental Management*, 43(5). 817–835. doi: 10.1007/s00267-008-9189-0.

Eriksen, S. and O'Brien, K. L., 2007. Vulnerability, poverty and the need for sustainable adaptation measures. *Climate Policy*, 7(4). 337–352.

Gbadegesin, A. S., Olorunfemi, F. B. and Raheem, U. A., 2010. Urban vulnerability to climate change and natural hazards in Nigeria. *Coping with Global Environmental Change, Disasters and Security – Threats, Challenges, Vulnerabilities and Risks*, H. G. Brauch, U. O. Spring, C. Mesjasz, J. Grin, P. Kameri-Mbote, B. Chourou, P. Dunay and J. Birkmann (eds), Hexagon Book Series on Human and Environmental Security and Peace, Vol. 5. Springer-Verlag, Berlin, 669–688.

Hauser, A., 2005. *Informe geológico geotécnico, preliminar: sectores Agüita de la Perdiz y Cerro La Pólvora, Concepción*, VIII Región. SERNAGEOMIN. Santiago, Chile, 14 p.

Huq, S. and Reid, H., 2007. *Community-Based Adaptation. A Vital Approach to the Threat Climate Change Poses to the Poor*. IIED Briefing Papers, IIED, London.

Ijaiya, G. T. and Umar, A. C., 2004. The informal and formal sector inter-linkages and the incidence of poverty in Nigeria: a case study of Ilorin Metropolis. *Africa Development*, 29(3). 84–102.

IPCC (Intergovernmental Panel on Climate Change), 2007. Climate change 2007: impacts, adaptation and vulnerability. *Contribution of Working Group II to the Fourth Assessment Report of the Intergovernmental Panel on Climate Change*, M. L. Parry, O. F. Canziani, J. P. Palutikof, P. J. van der Linden and C. E. Hanson (eds). Cambridge University Press, Cambridge, UK. 976 p.

Iversen, T., Benestad, R., Haugen, J. E., Kirkevåg, A., Sorteberg, A., Debernard, J., Grønås, S., Hanssen-Bauer, I., Kvamstø, N. G., Martinsen, E. A. and Engen-Skaugen, T., 2005. *Norges klima om 100 år. Usikkerheter og risiko*. (Norwegian Climate in 100 years. Uncertainty and risk.) RegClim, Oslo. http://regclim.met.no/presse/download/regclim_brosjyre2005.pdf.

Klein, R. J. T., Eriksen, S., Næss, L. O., Hammill, A., Robledo, C. and O'Brien, K., 2007. Portfolio screening to support the mainstreaming of adaptation to climate change into development. *Climatic Change*, 84(1). 23–44. doi: 10.1007/s10584-007-9268-x.

Leary, N. A., 1999. A framework for benefit–cost analysis of adaptation to climate change and climate vulnerability. *Mitigation and Adaptation Strategies for Global Change*, 4(3–4). 307–318. doi: 10.1023/A:1009667706027.

Leichenko, R. M. and O'Brien, K. L., 2008. *Environmental Change and Globalization: Double Exposures*. Oxford University Press, New York, USA.

Lowe, A., Foster, J. and Winkelman, S., 2009. *Ask the Climate Question: Adapting to Climate Change Impacts in Urban Regions*. Center for Clean Air Policy, Washington, DC, USA.

Mardones, M. and Vidal, C., 2001. La zonificación y evaluación de los riesgos naturales de tipo geomorfológico: un instrumento para la planificación urbana en la ciudad de Concepción. *EURE*, 27(81). 97–122. doi: 10.4067/S0250-71612001008100006.

Matthews, H. D. and Caldeira, K., 2008. Stabilizing climate requires near-zero emissions. *Geophysical Research Letters*, 35. L04705. doi: 10.1029/2007GL032388.

Mehrotra, S., Natenzon, C. E., Omojola, A., Folorunsho, R., Gilbride, J. and Rosenzweig, C., 2009. Framework for city climate risk assessment: Buenos Aires, Delhi, Lagos, and New York. *Proceedings of the World Bank Fifth Urban Research Symposium on Cities and Climate Change: Responding to an Urgent Agenda*, Marseille, France, 5–8 June.

Munasinghe, M. and Swart, R. (eds), 2000. *Climate Change and its Linkages with Development, Equity and Sustainability*. Intergovernmental Panel on Climate Change (IPCC), Geneva, Switzerland.

Næss, L. O., Bang, G., Eriksen, S. and Vevatne, J., 2005. Institutional adaptation to climate change: flood responses at the municipal level in Norway. *Global Environmental Change*, 15(2). 125–138. doi: 10.1016/j.gloenvcha.2004.10.003.

Nelson, D. R., Adger, W. N. and Brown, K., 2007. Adaptation to environmental change: contributions of a resilience framework. *Annual Review of Environment and Resources*, 32. 395–419. doi: 10.1146/annurev.energy.32.051807.090348.

O'Brien, K. L. 2009. Do values subjectively define the limits to climate change adaptation? *Adapting to Climate Change. Values, Thresholds, Governance*, W. N. Adger, I. Lorenzoni and K. L. O'Brien (eds). Cambridge University Press, Cambridge, UK. 164–180.

O'Brien, K. L., Eriksen, S., Sygna, L. and Næss, L. O., 2006. Questioning European complacency: climate change impacts, vulnerability and adaptation in Norway. *Ambio*, 35(2). 16–22.

O'Brien, K. L., St.Clair, A. and Kristoffersen, B., 2010. *Climate Change, Ethics and Human Security*. Cambridge University Press, Cambridge.

O'Brien, K. L., Sygna, L., Leichenko, R., Adger, W. N., Barnett, J., Mitchell, T., Schipper, L., Tanner, T., Vogel, C. and Mortreux, C., 2008. *Disaster Risk Reduction, Climate Change Adaptation and Human Security. A Commissioned Report for the Norwegian Ministry of Foreign Affairs*. GECHS Report 2008:3, University of Oslo, Norway.

Olorunfemi, F. B., 2008. Disaster incidence and management in Nigeria. *Research Review*, 24(2). 1–23.

Olorunfemi, F. B. and Raheem, U. A., 2007. Urban development and environmental implications: The challenge of urban sustainability in Nigeria. *The Australasian Review of African Studies*, 28(2006/2007). 74–96.

Olsson, P. and Folke, C., 2001. Local ecological knowledge and institutional dynamics for ecosystem management: a study of Lake Racken Watershed, Sweden. *Ecosystems*, 4(2). 85–104.

ONEMI (Oficina Nacional de Emergencia, Ministerio del Interior), 2005. Informe consolidado. Sistemas frontales sucesivos: 10 Mayo–15 Julio 2005. Ministerio del Interior, Santiago, Chile. 30 p.

Parry, M., Arnell, N., Berry, P., Dodman, D., Fankhauser, S., Hope, C., Kovats, S., Nicholls, R., Satterthwaite, D., Tiffin, R. and Wheeler, T., 2009. *Assessing the Costs of Adaptation to Climate Change. A Review of the UNFCCC and Other Recent Estimates*. International Institute for Environment and Development (UK), and the Grantham Institute for Climate Change, Imperial College London (UK), London.

Pelling, M. and High, M., 2005. Understanding adaptation: what can social capital offer assessments of adaptive capacity? *Global Environmental Change*, 15(4). 308–319. doi: 10.1016/j.gloenvcha.2005.02.001.

Reid, P. and Vogel, C., 2006. Living and responding to multiple stressors in South Africa – glimpses from KwaZulu-Natal. *Global Environmental Change*, 16(2). 195–206. doi: 10.1016/j.gloenvcha.2006.01.003.

Roberts, D., 2008. Thinking globally, acting locally – institutionalizing climate change at the local government level in Durban, South Africa. *Environment and Urbanization*, 20(2). 521–537. doi: 10.1177/0956247808096126.

Robinson, J. and Herbert, D., 2001. Integrating climate change and sustainable development. *International Journal on Global Environmental Issues*, 1(2). 130–149.

Sanchez-Rodriguez, R., 2009. Learning to adapt to climate change in urban areas. A review of recent contributions. *Current Opinion in Environmental Sustainability*, 1. 201–206. doi: 10.1016/j.cosust.2009.10.005.

Schellnhuber, H. J., 2009. Global warming: Stop worrying, start panicking? *PNAS*, 105(38). 14239–14240.

Schipper, E. L. and Pelling, M., 2006. Disaster risk, climate change and international development: Scope and challenges for integration. *Disasters*, 30(1). 19–38.

Smit, B., Burton, I., Klein, R. J. T. and Wandel, J., 2000. An anatomy of adaptation to climate change and variability. *Climatic Change*, 45(1). 223-251.

Storbjørk, S., Lähteenmäki-Smith, K. and Hilding-Rydevik, T., 2009. Conflict or consensus: The challenge of integrating environmental sustainability into regional development programming. *European Journal of Spatial Development*, 34. 1–22.

Tschakert, P. and Dietrich, K. A., 2010. Anticipatory learning for climate change adaptation and resilience. *Ecology and Society*, 15(2). 11. www.ecologyandsociety.org/vol15/iss2/art11/.

Ulsrud, K., Sygna, L. and O'Brien, K. L., 2008. *More than Rain: Identifying Sustainable Pathways for Climate Adaptation and Poverty Reduction*. Report prepared for the Development Fund, Norway.

Vaage, O. F., 2004, *Trening, mosjon og friluftsliv. Resultater fra Levekårsundersøkelsen 2001 og Tidsbruksundersøkelsen 2000. (Exercise, leisure and outdoors activities. Results from Living Conditions Survey 2001 and Time Use Survey 2000.)* Statistics Norway, Oslo/Kongsvinger. www.ssb.no/emner/07/02/50/rapp_200413/rapp_200413.pdf.

van Aalst, M. K., Cannon, T. and Burton, I., 2008. Community level adaptation to climate change: The potential role of participatory community risk assessment. *Global Environmental Change*, 18(1). 165–179. doi: 10.1016/j.gloenvcha.2007.06.002.

Vogel, C., Moser, S. C., Kasperson, R. E. and Dabelko, G. D., 2007. Linking vulnerability, adaptation, and resilience science to practice: Pathways, players, and partnerships. *Global Environmental Change*, 17(3–4). 349–364.

Winkler, H. and Marquand, A., 2009. Changing development paths: From an energy-intensive to low-carbon economy in South Africa. *Climate and Development*, 1. 47–65. doi: 10.3763/cdev.2009.0003.

Wisner, B., Blaikie, P., Cannon, T. and Davis, I., 2004. *At Risk: Natural Hazards, People's Vulnerability and Disasters*. Routledge, London, UK.

Woolcock, M. and Narayan, D., 2000. Social capital: implications for development theory, research, and policy. *World Bank Research Observer*, 15(2). 225–249.

Yohe, G. W., Lasco, R. D., Ahmad, Q. K., Arnell, N. W., Cohen, S. J., Hope, C., Janetos, A. C. and Perez, R. T., 2007. Perspectives on climate change and sustainability. *Climate Change 2007: Impacts, Adaptation and Vulnerability. Contribution of Working Group II to the Fourth Assessment Report of the Intergovernmental Panel on Climate Change*, M. L. Parry, O. F. Canziani, J. P. Palutikof, P. J. van der Linden and C. E. Hanson (eds). Cambridge University Press, Cambridge, UK. 811–841.

Ziervogel, G., Bharwani, S. and Downing, T. E., 2006. Adapting to climate variability: Pumpkins, people and policy. *Natural Resources Forum*, 30(4). 294–305. doi: 10.1111/j.1477-8947.2006.00121.x.

review article

climate and development

Sustainable adaptation: An oxymoron?

KATRINA BROWN*

School of International Development and Tyndall Centre for Climate Change Research, University of East Anglia, Norwich NR4 7TJ, UK

This article provides critical reflection on the notion of sustainable adaptation. It discusses how the concept resonates with debates on sustainable development, and how the key elements of sustainable adaptation presented in the literature relate to sustainable development. It discusses three major challenges to the promotion and implementation of sustainable adaptation and achievement of its objectives. First, many current approaches to adaptation are far from sustainable. Second, the relationships between poverty reduction and adaptation to climate change are complex and highly context specific. Third, sustainable adaptation may be co-opted to support development-as-usual rather than more radical options which put social justice, equity and environmental sustainability at the core.

Keywords: adaptation; maladaptation; poverty; sustainable development

1. Introduction

This article intends to provide a critical reflection and comment to supplement, and in some ways challenge, the other articles included in this special issue. It interrogates some of the implicit assumptions and complexities inherent in current discussions and promotion of the new concept of sustainable adaptation. The title of the article is both deliberatively provocative, in that it throws out a rather bold challenge to the concept of sustainable adaptation, and purposefully redolent of earlier debates about sustainable development, widely rehearsed more than two decades ago (notably Redclift, 1987; see also Redclift, 2006). These earlier discussions highlighted some of the problems with the concept of sustainable development. They observed that it is deliberately vague and slippery, making it extremely difficult to operationalize; that it enables 'greenwashing' or green camouflaging (Adams, 2001) of current strategies and thus in some way fosters hypocrisy; that it implies the notion and possibility of sustainable growth (what Daly (1990) referred to as an impossibility theorem); and that it is delusional, in distracting attention away from meaningful, more profound change and the root causes of global inequities and environmental degradation (Robinson, 2004). A substantial literature has been established debating these and many other issues. This literature encompasses the broad set of ideas and the wide range of different perspectives, which in fact make up sustainable development.

This article shows how sustainable adaptation shares some similar features with sustainable development. Sustainable adaptation, as a term and a concept, is now gaining currency in debates about how best to respond to climate change, particularly in poor countries. It highlights the types of interventions which might simultaneously address both climate change and poverty and development concerns. Although there is no agreed definition of sustainable adaptation, the following section explores some of the ideas and meanings that underpin the concept. As a working definition, this article takes sustainable adaptation to mean those

* *E-mail:* k.brown@uea.ac.uk

 doi:10.3763/cdev.2010.0062

ISSN: 1756-5529 (print)

adaptation measures that also contribute towards social equity and environmental integrity, two pillars of sustainable development. But as the articles in this special issue and the emerging literature show, there is no one agreed definition. Thus, like sustainable development, sustainable adaptation might be viewed as 'slippery'. The concept of sustainable adaptation is similar to sustainable development in two further important aspects. First is the idea that there can be synergy between at least two objectives which are often seen as conflicting. Second, that there are ways in which these objectives can be integrated in policy prescriptions and specific interventions on the ground. In other words, these two processes – synergy and integration – are also implied in discussions of sustainable adaptation.

The next section reviews the concept of sustainable adaptation and some of the assumptions behind it. It highlights some of the contradictions, tensions and ambiguities in the term and its applications, which in some ways resonate with earlier debates about sustainable development. As with sustainable development, these very features also make the concept attractive – to scientists and policymakers alike. The article then outlines three areas where these complications are played out and identifies some of the problems faced in finding a shared understanding, constructive implementations and effective achievement of the objectives of sustainable adaptation. The first issue concerns the way in which adaptation is currently approached and implemented. Adaptation is often seen as comprising actions designed and led from outside. Analysis shows that current adaptation interventions are often not sustainable; they may undermine long-term resilience and may even result in 'maladaptation'. Second, a lack of clarity about the relationship between adaptation to climate change and poverty alleviation means that finding synergistic actions is much more difficult than often assumed. Third, sustainable adaptation ultimately challenges us to identify and address the root causes of both environmental change and poverty, which may require re-thinking current dominant ideas and practice in development. These issues can potentially hinder the wider promotion and adoption of sustainable adaptation and provide obstacles to the successful achievement of its objectives.

2. What is sustainable adaptation?

Ideas about sustainable adaptation have recently emerged as part of a wider discourse which seeks to situate concerns about climate change and vulnerability in a context of sustainable development, and thereby to more closely align interventions to support adaptation to climate change with risk reduction and poverty alleviation. Thus, Ulstrud et al.'s (2008) recent report for the Development Fund in Norway talks of 'identifying sustainable pathways for climate change adaptation and poverty reduction'. O'Brien and Leichenko (2007, p. 31) provide a definition of sustainable adaptation, such that 'combining aspects of both sustainability and adaptation, the notion of sustainable adaptation entails measures that reduce vulnerability and promote long-term resilience in a changing climate'. Eriksen and O'Brien (2007, p. 341) present sustainable adaptation as a set of responses which form the overlap between poverty reduction measures on the one hand, and vulnerability reduction measures on the other. Eriksen and O'Brien present a thorough analysis of the links between poverty and vulnerability to climate change, arguing that sustainable adaptation interventions must focus on (1) reducing risks to current ways of securing well-being, (2) strengthening adaptive capacity of the poor and (3) addressing the causes of vulnerability among the poor. A sustainable adaptation measure may focus on any one of these linkages, as long as it does not negatively affect the other two (Eriksen and O'Brien, 2007, p. 342).

These definitions imply that sustainable adaptation brings two extra dimensions – or two additional elements or requirements – to conventional adaptation approaches: first it should actively reduce poverty and/or vulnerability, and second, it must ensure that actions are sustainable in the long term and do not undermine resilience, that is, that it includes measures which do not compromise longer-term sustainable development or ability to respond to climate change and other stressors. These two aspects of sustainable adaptation echo the intra- and inter-generational dimensions of sustainable development.

Sustainable adaptation can be seen as part of a broader family of approaches seeking to combine the objectives of climate change adaptation and poverty alleviation. Thus, there are obvious linkages and similarities with ideas around pro-poor climate change adaptation (Tanner and Mitchell, 2008a), community-based adaptation (Ensor and Berger, 2009) and climate change adaptation, which actively helps to reduce poverty (e.g. Tanner and Mitchell, 2008b). These proposals show how, after a dominant view that perceived measures to address climate change as a threat to development and which primarily emphasized its potential negative impacts on the poor, the focus has now turned to finding approaches that ensure the compatibility of climate change adaptation and poverty alleviation (Adger et al., 2003). Significantly, there is still much less written about the possibility of pro-poor climate change mitigation than about adaptation (most discussions have focused on carbon forestry and Clean Development Mechanism (CDM), potential Reducing Emissions from Deforestation and Forest Degradation (REDD) and low carbon development – see Yohe et al., 2007; Peskett and Prowse, 2008; Urban, 2010).

3. How sustainable is current adaptation?

The framings of sustainable adaptation outlined above beg a number of questions: is sustainable adaptation significantly different from current adaptation? Is current adaptation sustainable? What can we learn from current adaptation to inform sustainable adaptation? First, it is important to recognize that adaptation itself is very broadly interpreted. Adaptation to climate change has emerged only in the past decade as a core issue in international development (Adger et al., 2003; Parry, 2009). Since then we have seen multiple experiments with adaptation around the world. McGray et al. (2007) provide an authoritative survey and assessment of 135 adaptation cases, revealing the range of approaches undertaken in developing countries. They present a spectrum of approaches to adaptation that moves from those most closely allied with conventional development and which address the drivers of poverty, to those that build capacity to respond, to a focus on managing specific climate risks, through to confronting specific climate change impacts. This has been presented as a continuum from a focus on vulnerability to a focus on impacts. Ensor and Berger (2009) categorize adaptation approaches somewhat differently, but again on a continuum. In their schema, vulnerability reduction targets a particular hazard. Strengthening of resilience reduces vulnerability to a wider set of hazards, and is supported by attempts to diversify assets or livelihoods. Building adaptive capacity both strengthens resilience and reduces vulnerability to a wide range of hazards, while addressing the amount, diversity and distribution of assets and actively facilitating alternative strategies. They refer to the 'adaptation space' as determining which strategy is appropriate in any given situation; the adaptation space itself is shaped by the relationship between uncertainty and vulnerability.

These and other reviews of current adaptation measures demonstrate that only a minority conform to sustainable adaptation as defined above. Furthermore, other studies indicate that many responses to climate change may in fact be far from sustainable and may be detrimental to the poor, or to the environment. Carr (2008) found that adaptation re-enforces and legitimates gender roles in Ghana, and Eriksen and Lind (2009) show how adaptation may strengthen or retain existing power structures and inequalities and exacerbate conflict in Kenya. Turner et al. (2010) demonstrate how adaptation may undermine biodiversity conservation. Livelihoods diversification is seen as a primary means of lifting people out of poverty (Ellis, 2000; Barrett and Swallow, 2006) and was also found to be the most widely cited autonomous adaptation strategy among smallholder farmers in Africa in a recent study by International Food Policy Research Institute (IFPRI) (Below et al., 2010). However, the study points out that diversification also brings increased and new risks as a result of market integration. It may also be environmentally unsustainable (e.g. diversifying into activities such as charcoal production or artisanal mining) and may raise inequality both within and between households.

Barnett and O'Neill (2010) provide an analysis of these maladaptations.[1] Their study identifies five pathways to maladaptation that could be used to screen adaptation decisions for their potential detrimental impacts. These five pathways identify maladaptation as actions which increase emissions of greenhouse gases; disproportionately burden the most vulnerable; have high opportunity costs; reduce incentives to adapt; or create path dependency. They also highlight that a critical problem dogging adaptation is the time lag between the changes in climate and the changes in institutions, which often masks the longer term and geographically and socially distant impacts of adaptation responses.

An analysis of nine climate change responses by Adger et al. (2010) identifies three factors which influence how the responses affect short-term coping versus long-term sustainability. Table 1 summarizes Adger's analysis of these nine responses. Narrow, technological perspectives allow only a limited number of possible responses to identified threats. Conversely, broader more inclusive frames recognize the importance of other system drivers and of maintaining response flexibility and tend to focus on the management of problems rather than on the implementation of specific actions. The policy responses to pine bark beetle infestation in western Canada and declining fish stocks in Uganda are examples of narrow, technological approaches. In western Canada, activities are designed to maximize short-term economic output of forest resources. In Uganda, government action is directed at controlling fishing technologies. In contrast, water resources management in Brazil moves beyond a supply and demand approach to address the multiple development and economic drivers of water use and value in basins. Responses in the Cayman Islands focus on tropical storm preparedness, and also incorporate an active learning cycle and wider governance issues.

Formal governance arrangements include vertical, top-down approaches as well as decentralized and more horizontally integrated systems. Vertical top-down governance structures are usually associated with government-led and managed initiatives. Drought management in north-east Brazil is an example of a highly vertical approach that has been detrimental to the emergence of more local and expansive institutions that represent adaptive capacity. These top-down actions are often associated with narrow technologically defined problems. More broadly considered responses, such as coastal management in the UK, tend to be more open to participation and contributions from a wider set of stakeholders at different scales. Ideally, there is an overlap between actors who define problems and responses and those who are affected by the response. However, there are often ramifications at temporal and spatial scales outside those at which decisions are made. Biofuels promotion, for example, is linked with increased deforestation in the Amazon and changes in land cover and soil conservation management in the United States.

The effects of each response may be displaced over time or space. Social learning is more likely if feedbacks occur soon relative to action and if those most affected by feedbacks are those responsible for the action. Slow feedbacks, those that are spatially distant, or those that are masked or ignored because of a focus on short-term economic, political or productivity gains, are less likely to affect policy responses. With respect to biofuels, ecological impacts are overlooked as a result of short-term economic gain and are dislocated through space.

These findings reiterate lessons from other studies: if climate change responses are to avoid maladaptation, climate change impacts need to

TABLE 1 Nine examples of observed responses to climate change-related resource stresses

Response to climate change	Problem framing	Governance structure	Ability to deal with feedbacks
Maize-based biofuel production in north America for greenhouse gas mitigation	Narrow – technological fix	Top-down	Low – impacts dislocated in time and space
Drought hazard compensation policy to cope with increased aridity in north-east Brazil	Narrow – focus on one issue	Top down government centred	Low – poor ability of institutions to respond
Public and private adaptations to variable rainfall in Canadian prairie agro-ecosystems	Narrow – single issue focused	Government led, involving public and private actors	Low
Government-led planning for tropical storm and hurricane impacts, Cayman Islands	Broad – looks at range of learning and institutional possibilities affecting disaster preparedness	Government led – inclusion of wider stakeholders	Medium
Forest sector response to pine beetle infestation, western Canada	Narrow – single sector focused (crop damage)	Top-down, forest-sector led	Low
Public and private response to drought and flood incidence in semi-arid Kenya	Narrow – single sector focused (land degradation) rather than linked livelihoods and rangelands	Top-down, public and private	Low – local knowledge undermined
Decentralized management of water quality and supply to cope with drought in Brazil	Broad – considers economic and development drivers relating to water quality and supply	Inclusionary process of stakeholder involvement, decentralized	Medium
Public sector planning to manage risks of increased coastal erosion and inundation, UK	Medium	Government and civil society – cross-scale public sector	Medium – local sources of knowledge can inform national policy
Public and private response to fluctuating lake levels and fish catches associated with variable rainfall and other stressors, Uganda	Narrow – focus on managing fish stocks rather than livelihood systems and lake ecosystem	Cross-scales and broad stakeholders, public and private with strong legacy of top-down regulation	Low – impacts felt locally but individual behaviour masked by short-term economic gains, pressing livelihood needs, failure to act collectively

Source: Adger et al. (2010).

be seen within a broader context of multiple stressors; to address multi-scale processes through diverse institutions and in partnership with multiple stakeholders; and to adopt flexible learning approaches. These lessons could provide preliminary principles to guide sustainable adaptation. However, they may be necessary, yet not sufficient, conditions for its success.

4. Clarifying relationships between adaptation and poverty reduction

As shown above, reduction of poverty and vulnerability through adaptation to climate change is a key component of sustainable adaptation. This section explores why it may be more difficult to find synergy between climate change adaptation and poverty reduction than is often assumed in literature and policy. Much of the existing literature – and policy – is based on the assumption that alleviating poverty will enable people to adapt to climate change; in other words, that there is an almost complete overlap between those who are poor and those who are vulnerable to the impacts of climate change. Eriksen and O'Brien (2007) show why this is not the case, highlighting how the experiences and states of vulnerability and poverty are distinct. They observe that many of the processes which generate vulnerability to climate change are closely associated with poverty, but that there is no direct mapping of climate change vulnerability and poverty. Not all poor people are vulnerable to climate change and certainly not all in the same ways, and not all of those vulnerable to climate change are indeed poor. Brouwer and Nhassengo (2006), for example, document how wealthier people were more affected by floods in Mozambique in terms of loss of assets. Eriksen and O'Brien cite Coetzee:

> Poverty and vulnerability do not coincide in the same way in all cases. People experiencing vulnerability are not necessarily poor: and amongst the poor, there may be varying levels and patterns of vulnerability – depending on the multitude of dynamic processes through which individuals and households respond to changes in the environment, adopt and adjust strategies and re-configure their relative well-being (Coetzee, 2002, p. 5, cited in Eriksen and O'Brien, 2007, p. 340).

The vulnerability or security of individuals and of societies is determined not only by the direct impacts of climate change on the resources upon which people depend, but also by the availability of resources and peoples' entitlements to call on them. Although poverty and marginalization are key driving forces of vulnerability and constrain individuals and households in their coping and long-term adaptation, vulnerability to future climate change is likely to have distinct new patterns and characteristics. It is highly likely that those people who are currently marginalized will be among the more severely affected, and indeed it has been argued that climate change is likely to exacerbate and reinforce inequalities both through dynamic vulnerability and the processes of adaptation itself (Adger et al., 2003).

This implies that the process of adaptation itself may create inequities or impoverish those who are already worse off. In this regard, it is important to distinguish between the different types of adaptation, not only anticipatory and reactionary, but also between coping (i.e. short-term measures which may undermine long-term capacity (Eriksen et al., 2005)) and adapting, and between autonomous and planned adaptation. Adaptation is often presented as a series of planned interventions, discrete events initiated from outside or by governments, when more usually individuals and households undertake spontaneous incremental and continuous adaptation, integrated with and balanced against other livelihood adjustments in the face of multiple stressors (Leichenko and O'Brien, 2008).

Accepting that both adaptation and poverty are complex, multi-dimensional and dynamic implies that there are no one-size-fits-all solutions; that sustainable adaptation is a moving target, meaning different things in different contexts. Eriksen and O'Brien (2007) stress this point. Tanner and Mitchell (2008b, p. 10) distinguish between the chronic poor, transient poor and the non-poor (and sub-categories of these), and point to the types of adaptation measure which these groups are most likely to undertake (see Table 2). For example, autonomous adaptations undertaken by the 'always poor' (which might be classified as coping strategies), such as selling assets, sending younger children to work or moving to more exposed areas, may make them more vulnerable and less able to cope with future problems, and therefore cannot be viewed as sustainable adaptation (see top left-hand box in Table 2). Autonomous adaptation by the transient poor – such as diversifying livelihoods – may aid long-term adaptation and reduce poverty, or it may undermine longer-term resilience or exposure to new risks. Disaggregating poverty in this way can be helpful to inform and qualify where, when and what kinds of sustainable adaptation may be possible.

TABLE 2 Adaptation options by poverty category

Aggregate category	Chronic poor		Transient poor		Non-poor
Specific poverty category	Always poor	Usually poor	Cyclical poor	Occasionally poor	Never poor
Autonomous adaptation	Selling of last assets	Intra-community transfer/charity	(Seasonal) migration	Diversify livelihoods	Investments in multiple financial assets
	Sending younger children to work	Sending children to work	Less risky production	Investment in social capital/ assets	Buy drought-tolerant seeds, new technologies
	Conflict, crime, sex work	Migration	Working multiple jobs, longer hours		Diversify livelihoods
	Move to more exposed locations	Extended family	Investment in social capital/reciprocity		Invest in ethical/green options
	Use fragile ecological assets	Conflict, crime, sex work			
Market-based adaptation		Promote micro-savings, micro-credit, micro-insurance	Weather-indexed insurance	Weather-indexed insurance	Crop insurance
		Cattle insurance	Cattle insurance	Promote micro-savings, micro-credit, micro-insurance	Farm asset insurance and domestic insurances
			Promote micro-savings, micro-credit, micro-insurance Selling assets	Selling assets	Price hedging
Public policy-driven adaptation	Social pensions	Community restocking	Community restocking	Improved remittances schemes	Taxation to pay for adaptation of the poor
	Assisted migration	Subsidized seed banks	Improved remittances schemes	Improved climate information	Marketing of green agenda
	Democratization of natural resource management/ ecosystem rehabilitation	Democratization of natural resource management/ ecosystem rehabilitation	Subsidized seed banks	Employment assurance schemes	Incentives for adaptation and mitigation choices
	Promote health and nutrition services	Promote health, disease reduction	Improved climate information	Social insurance schemes	
	Cash for work schemes	Cash for work schemes	Ecosystem rehabilitation	Irrigation schemes and urban services provision	
	Conditional cash transfers		Irrigation schemes		

Source: Tanner and Mitchell (2008b, p. 10).

Only some of these myriad responses can be construed as sustainable adaptation. Furthermore, some of them imply important trade-offs. For example, the promotion of market-based approaches, while building people's assets and reducing poverty and perhaps reducing vulnerability to one type of impact, may make households vulnerable to another set of shocks, for example, price fluctuations or market shifts (see Peskett and Prowse (2008), for discussion of biofuels). This underscores that it is important to understand the multiple dimensions and dynamics of poverty and the systemic causes of vulnerability. It also exposes how development policies which address certain aspects of poverty may have detrimental impacts on response capacity and sustainable adaptation. Bunce et al. (2010) show how national and regional development and conservation policies undermine the resilience of poor coastal communities in East Africa. Likewise Eriksen and Silva (2009) (see also Silva et al., 2010) show how shifts to commercial agriculture, land privatization and market integration – all associated with conventional rural development and pathways out of poverty – can weaken the ability of smallholder farmers to respond to drought. Barnett (2008) shows that, in the South Pacific, dependence on development aid militates against autonomous anticipatory action and decreases adaptive capacity by undermining the legitimacy of government action, and ultimately makes sustainable adaptation (or sustainable development) much less likely. In each of these cases, conventional development interventions are shown to make poor people more vulnerable to climate change and less able to adapt.

5. Do we need to re-think development for sustainable adaptation?

Until relatively recently, many approaches to climate change adaptation assumed that conventional poverty reduction strategies were the best way to alleviate climate change impacts and to lessen vulnerability. Indeed, much adaptation research emphasizes the need to mainstream adaptation into current development strategies (e.g. poverty reduction strategy papers), and considers the implications of adaptation for current aid flows and development portfolios (Klein et al., 2007; Huq and Ayers, 2008; Prowse and Grist, 2009). But these approaches do not question the underlying direction and type of development. In fact, Grist's (2008) analysis shows that current approaches to climate change adaptation are aligned with reformist discourses of sustainable development, notably market environmentalism, ecological modernization and environmental populism and not more radical interpretations. This business-as-usual approach to climate change adaptation in development policy is exemplified by 'climate change screening', an emerging approach that aims to embed climate change into development planning. The approach involves assessing existing development plans are for their exposure to climate change risks, and identifying measures to reduce these risks (Klein et al., 2007). These adaptation discourses thus emphasize the necessity to control and minimize change and have as their ultimate objective the maintenance of the status-quo or business-as-usual. This managerial approach to adaptation stresses the more effective deployment of existing institutional, technological and market-based risk reduction measures such as disaster preparedness, irrigation and insurance (e.g. UNFCCC, 2007). In particular, this approach fails to address the risks associated with large, severe and abrupt climatic and environmental changes. But more importantly, it does not address the root causes or underlying processes which actually make people both vulnerable to climate change, which again makes sustainable adaptation less likely.

Therefore, adaptation as currently conceived is firmly embedded within the dominant development paradigm and effectively treats climate change as an externality. This is exemplified by the World Bank's policies for 'climate resilient development' which seek to protect economic growth from the ravages of climate change (Brown, 2010). Economic growth, modernization and markets are still viewed as the core drivers of development and of poverty alleviation. Thus, climate change is viewed by most development practitioners as something that affects development, and from which development needs to be protected, rather than a symptom of unsustainable development. Yet Jerneck and Olsson (2008) argue that development, as a modernization project, offers few solutions for either adaptation or sustainability. Similarly, Brooks et al. (2009) conclude that current development models and adaptation approaches are inadequate for addressing climate change. Through their focus on growth and consumption and their treatment of the environment as an externality, existing development models drive climate change and can exacerbate vulnerability to it. While existing approaches to adaptation may go some way towards reducing the risks climate change poses to development investments, these approaches run the risk of locking societies into unsustainable and maladaptive patterns of development, by assuming that development along business-as-usual lines can simply be secured through the identification and implementation of appropriate adaptation measures. Although Cannon and Muller-Mahn (2010) contend that climate change adaptation may already be changing the development discourse, fundamental issues around power inequality, the primary cause of poverty and climate change, are ignored. Without putting such issues at the heart of policy, neither development nor adaptation can be truly 'pro-poor'. This implies that for sustainable adaptation to really meet its objectives, it must more radically challenge current models of development and adaptation and present workable alternatives.

6. Conclusion: is sustainable adaptation an oxymoron?

Sustainable adaptation brings extremely important contributions to discussions about climate change and development. It allies climate change adaptation closely to sustainable development and highlights the importance of recognizing the interactions and links between poverty and social justice, and environmental sustainability, when considering responses to climate change. It requires that adaptation to climate change is considered within the context of multiple stressors and vulnerabilities, and with concern for both intra- and inter-generational equity. However, in doing so, sustainable adaptation potentially risks falling into similar traps as sustainable development has done. The term itself is rather vague and open to a wide range of interpretations. It is difficult to translate into policy, and there is a danger it will be used to justify existing policies and approaches that are not sustainable. Ultimately, climate change requires us to challenge and re-configure ideas about development and to consider alternative strategies, including low carbon and degrowth of the economy. Sustainable adaptation is an oxymoron unless it specifically deals with fundamental problems in the dominant paradigm of unsustainable development. In particular, it needs to examine its approach to multi-dimensional poverty reduction and the need for fundamental institutional re-configuration in support of long-term equity and resilience.

Acknowledgements

This research was partly funded by the ESRC Professorial Fellowship 'Resilient Development in Social Ecological Systems'. An earlier version of this article was presented at the GECHS conference in June 2009, and I am grateful to participants there and to anonymous reviewers for insightful comments on earlier versions. All views and errors remain the responsibility of the author.

Note

1. Maladaptation is defined as 'action taken ostensibly to avoid or reduce vulnerability to climate change that impacts adversely on, or increased the vulnerability of other systems, sectors or social groups' Barnett and O'Neill (2010, p. 211).

References

Adams, W. M., 2001. *Green Development: Environment and Sustainability in the Third World* (2nd edn). Routledge, London.

Adger, W. N., Brown, K., Nelson, D. R., Berkes, F., Eakin, H., Folke, C., Galvin, K., Goulden, M., Gunderson, L., O'Brien, K., Ruitenbeek, J. and Tompkins, E. L., 2010. Resilience implications of responses to climate change. WIRESCC (Wiley Interdisciplinary Reviews in Environmental Science: Climate Change).

Adger, W. N., Huq, S., Brown, K., Conway, D. and Hulme, M., 2003. Adaptation to climate change in the developing world. *Progress in Development Studies*, 3(3). 179–195.

Barnett, J., 2008. The effect of aid on capacity to adapt to climate change: Insights from Niue. *Political Science*, 60(1). 31–45.

Barnett, J. and O'Neill, S., 2010. Maladaptation. *Global Environmental Change*, 20. 211–213.

Barrett, C. B. and Swallow, B. M., 2006. Fractal poverty traps. *World Development*, 34(1). 1–15.

Below, T., Artner, A., Siebert, R. and Sieber, S., 2010. *Micro-Level Practices to Adapt to Climate Change for African Small-Scale Farmers*. IFPRI Discussion Paper 00953, Environment and Production Technology Department, IFPRI.

Brooks, N., Brown, K. and Grist, N., 2009. Development futures in the context of climate change: challenging the present and learning from the past. *Development Policy Review*, 27(6). 741–765.

Brouwer, R. and Nhassengo, J., 2006. About bridges and bonds: community responses to the 2000 floods in Mabalane District, Mozambique. *Disasters*, 30(2). 234–255.

Brown, K., 2010. *Analysing Policy Discourses of Resilience*. Paper presented at the American Association of Geographers Conference, Washington, DC, USA, 14–18 April.

Bunce, M., Brown, K. and Rosendo, S., 2010. Policy misfits, climate change and cross-scale vulnerability in coastal Africa: how development projects undermine resilience. *Environmental Science and Policy*, 13. 485–497.

Cannon, T. and Muller-Mahn, D., 2010. Vulnerability, resilience and development discourses in context of climate change. *Natural Hazards*, 55. 621–635. doi: 10.1007/s11069-010-9499-4.

Carr, E. R., 2008. Between structure and agency: Livelihoods and adaptation in Ghana's central region. *Global Environmental Change*, 18. 689–699.

Coetzee, E., 2002. Urban vulnerability: A conceptual framework. *Urban Vulnerability: Perspectives from Southern Africa*, C. Nomdo and E. Coetzee (eds). Periperi Publications, Cape Town, South Africa.

Daly, H., 1990. Sustainable growth: An impossibility theorem. *Development*, 3/4. 45–47.

Ellis, F., 2000. *Rural Livelihoods and Diversity in Developing Countries*. Oxford University Press, Oxford.

Ensor, J. and Berger, R. 2009. *Understanding Climate Change Adaptation: Lessons from Community-Based Approaches*. Practical Action Publishing, London.

Eriksen, S., Brown, K. and Kelly, P. M., 2005. The dynamics of vulnerability: locating coping strategies in Kenya and Tanzania. *Geographical Journal*, 171(4). 287–305.

Eriksen, S. and Lind, J., 2009. Adaptation as a political process: adjusting to drought and conflict in Kenya's drylands. *Environmental Management*, 43. 817–835.

Eriksen, S. and O'Brien, K., 2007. Vulnerability, poverty and the need for sustainable adaptation measures. *Climate Policy*, 7. 337–352.

Eriksen, S. and Silva, J., 2009. The vulnerability context of a savanna area in Mozambique: household drought coping strategies and responses to economic change. *Environmental Science and Policy*, 12. 33–52.

Grist, N., 2008. Positioning climate change in sustainable development discourse. *Journal of International Development*, 20. 783–803.

Huq, S. and Ayers, J. 2008. *Streamlining Adaptation to Climate Change into Development Projects at the National and Local Level*. European Parliament, Brussels.

Jerneck, A. and Olsson, L., 2008. Adaptation and the poor: development, resilience and transition. *Climate Policy*, 8. 170–182.

Klein, R. J. T., Eriksen, S. E. H., Næss, L. O., Hammill, A., Robledo, C., O'Brien, K. L. and Tanner, T. M., 2007. *Portfolio Screening to Support the Mainstreaming of Adaptation to Climate Change into Development Assistance*. Tyndall Centre Working Paper, 102 University of East Anglia, UK.

Leichenko, R. M. and O'Brien, K. L., 2008. *Environmental Change and Globalisation: Double Exposures*. Oxford University Press, Oxford.

McGray, H., Hammill, A. and Bradley, R., 2007. *Weathering the Storm: Options for Framing Climate Change and Development*. WRI Report, World Resources Institute, Washington, DC.

O'Brien, K. and Leichenko, R., 2007. *Human Security, Vulnerability and Sustainable Adaptation*. Human

Development Report Office, Occasional Paper 2007/ 9. UNDP, New York.

Parry, M., 2009. Climate change is a development issue, and only sustainable development can confront the challenge. *Climate and Development*, 1. 5–9.

Peskett, L. and Prowse, M., 2008. *Mitigating Climate Change: What Impact on the Poor?* ODI Opinion, vol. 97, Overseas Development Institute, London.

Prowse, M. and Grist, N., 2009. *The Environment/Development Disconnect: A Case Study of NAPAs and PRSPs*. Project Briefing Note. Overseas Development Institute, London.

Redclift, M., 1987. *Sustainable Development: Exploring the Contradictions*. Routledge, London.

Redclift, M., 2006. Sustainable development (1987–2005) – an oxymoron comes of age. *Horizontes Anthropológicos*, 12(25). 65–84.

Robinson, J., 2004. Squaring the circle? Some thoughts on the idea of sustainable development. *Ecological Economics*, 48. 369–384.

Silva, J., Eroiksen, S. and Ombe, Z. A., 2010. Double exposure in Mozambique's Limpopo River Basin. *Geographical Journal*, 176(1). 6–24.

Tanner, T. and Mitchell, T., 2008a. Introduction: building the case for Pro-poor adaptation. *IDS Bulletin*, 39(4). 1–5.

Tanner, T. and Mitchell, T., 2008b. Entrenchment or enhancement: Could climate change adaptation help to reduce chronic poverty? *IDS Bulletin*, 39(4). 6–15.

Turner, W. R., Bradley, B. A., Estes, L. D., Hole, D. G., Oppenheimer, M. and Wilcove, D. S., 2010. Climate change: helping natural survive the human response. *Conservation Letters*, 3(5). 304–312.

Ulstrud, K., Sygna, L. and O'Brien, K., 2008. *More Than Rain: Indentifying Sustainable Pathways for Climate Adaptation and Poverty Reduction*. The Development Fund, Oslo.

UNFCCC, 2007. *Climate Change: Impacts, Vulnerabilities and Adaptation in Developing Countries*. UNFCCC Secretariat, Bonn.

Urban, F., 2010. The MDGs and beyond: can low carbon development be pro-poor? *IDS Bulletin*, 41(1). 92–99.

Yohe, G., Burton, I., Huq, S. and Rosegrant, M., 2007. *Climate Change: Pro-poor Adaptation, Risk Management and Mitigation Strategies*. 2020 Focus Brief, IFPRI, Washington, DC.

research article

climate and development

Converging and conflicting interests in adaptation to environmental change in central Vietnam

MALIN BECKMAN*

Department of Urban and Rural Development, Swedish University of Agriculture Sciences, Box 7012, 750 07 Uppsala, Sweden

The article is based on qualitative research in the provinces of Quang Tri and Thua Thien Hue in central Vietnam during the years 1996–2009, by the author and colleagues at Hue University of Agriculture and Forestry. The focus of the article is on policies that, while increasing resilience at one scale, may cause increased vulnerability at other scales. Policies on forest protection and construction of hydroelectric dams contribute to regulating flooding of the lowland areas. However, the policies also result in severe constraints in access to land and forest products for the mountain population, which has impacts on their capacity to manage risk and adapt to environmental change. Forest resources have previously functioned as an important buffer for mountain households when coping with crises like serious floods. This reduced adaptive capacity may be a critical issue for equity and social sustainability in adaptation.

Keywords: adaptation; adaptive capacity; climate change; equity; forest policies; social sustainability; Vietnam; vulnerability

1. Introduction

1.1. Context and purpose

There is growing attention in Vietnam to adaptation to climate change. In 2008, the Vietnamese government adopted a National Target Program for Climate Change. The discourse on how to respond to climate change includes maintaining and increasing forest cover, water management infrastructure, adaptation of crop species and changed production patterns. This article discusses how adaptation measures that are intended to increase resilience may simultaneously cause increased vulnerability at other scales. Policies on forest protection and construction of hydroelectric dams and reservoirs are not implemented particularly in response to climate change, but these policies are now increasingly being justified from the perspective of flood regulation and preventing salt water intrusion into lowland cultivated areas. Although forest protection and reservoirs generally benefit both lowland and mountain populations, such measures also result in severe constraints in access to forest and land resources for the mountain population. There are other examples of tension where development strategies have consequences for both resilience and vulnerability, as for example the Vietnam government's focus on rice production, intensification of production and commercialization. This article argues that this may reduce their capacity to adapt to environmental change. This is a critical issue for equity and social sustainability in adaptation.

Vietnam has strong advantages in responding to climate-related hazards because it is a highly organized society with traditions of community responsibility for common welfare. However, this research suggests that not all people benefit from society's resilience because of biases towards lowland interests in the country.

In Vietnam, climate change impacts are likely to be most severe for the coastal areas that are

■ **E-mail:* malin.beckman@sol.slu.se

ISSN: 1756-5529 (print)

exposed to sea-level rise, salt water intrusion and flooding. Salt water intrusion is already a problem in many river systems, reducing productivity of crops, reducing areas of cultivatable land and affecting groundwater systems for drinking water (Carew-Reid, 2008). However, climate variability and extreme floods also cause serious problems in the mountains, as the force of the water causes erosion, landslides and damage to crops in the river valleys.

Several studies provide evidence that central Vietnam is experiencing increasingly unpredictable weather, which could be connected to climate change (Hanh, 2010; Phuong, 2010). Floods and drought are appearing at unexpected times of the year, making it increasingly difficult to plan the seasonal calendar for agriculture production. People are experiencing weather extremes, such as the 1999 floods (the largest in memory) when it rained 2,300 mm in four days. The year 2008 was the coldest winter in memory, and many animals died. The following autumn in Thua Thien Hue province there were seven floods between September and November. Water did not have time to drain in between the floods, but gradually rose higher and higher, reaching the roof tops in some areas. February 2010 saw temperatures of up to 40°C, even though it was winter (Hanh, 2010; Phuong, 2010 and the author's own experience).

This article argues that there is a need for attention to the concept of adaptive capacity and the resources used by households to cope with increasing climate stress. Within this perspective it is argued that development strategies resulting in an alienation of poor people from the forests reduce the flexibility of resources required for adaptive capacity of poor forest-dependent communities.

The article is based on fieldwork that was conducted in Hai Lang district, Quang Tri province (lowland and hill areas) and in A Luoi district, Thua Thien Hue province (mountain areas) during the period 2000–2005, following the disastrous floods of 1999. Five villages were studied, with different geographical, socio-economic and institutional conditions. The fieldwork was mainly conducted as qualitative semi-structured interviews with 15 households in each village. The interviews covered issues of how the household coped with and recovered from the flood losses and their perspectives on vulnerability and resilience. The interviews were repeated three or four times during the study period. Group discussions were held with groups of around 10 households. Tools such as mapping, institutional diagrams and ranking were used to gather comments on relationships to organizations, support structures and respondents' perspectives on risk and how this affects their livelihood strategies. Interviews were held with staff of local organizations and local government at commune and district levels on their experiences in relation to the floods and the recovery process. Interviews were also held with staff members of Hue University of Agriculture and Forestry who work in these communes and districts with research and community development. During the years 2007–2010 the author spent four months per year in these provinces, teaching and supervising MSc students of rural development and guiding junior researchers as part of a cooperation programme with Hue University of Agriculture and Forestry. This meant that the research on the process of recovery from the 1999 floods could be followed up and discussed in light of the frequent climate stress of the past decade, as well as development measures and adaptation in the study area.

1.2. The concepts of adaptation and adaptive capacity

The Intergovernmental Panel on Climate Change (2007) defines adaptation as 'an adjustment in natural or human systems in response to actual or expected climatic stimuli or their effects, which moderates harm or exploit beneficial opportunities'. The main focus of the National Target Program of the Vietnam Government is on the planned means of adaptation. This article argues for a greater focus on the potential for autonomous adaptation, which is being undermined by the reduced access to forest land for the mountain population.

This article focuses on aspects of adaptation to climate change that involve the capacity to manage increasingly unstable, uncertain and extreme weather phenomena. It also discusses how communities manage the increased competition for land resources that may follow from sea-level rise and salt water intrusion.

The concept of adaptive capacity focuses on the need for continuous flexibility and risk management, rather than looking for 'solutions'. Thomalla et al. (2006) highlight the contrast between adaptation strategies in households, which turn to multiple sources of security, and the more sectoral and technical approaches promoted by national and international institutions. The fieldwork presented here shows that coping and recovery from disasters caused by flooding require access to a range of alternative resources in order to manage losses. Similar principles apply to adaptation, as the environment that people adapt to is one of frequent disasters and high-production risk.

Nelson et al. (2007) discuss adaptive capacity in terms of the characteristics that facilitate flexibility in resource use, such as diversity of livelihood options and decision-making structures, which enable adaptation to change and shocks. Their focus is thus on preconditions, which increase the ability of individuals, groups and organizations to adjust to and deal with change. This article builds on their thinking in discussing how policies for adaptation lead to changes in the preconditions vital for risk management and adaptation, particularly for the mountain population in this study. Change to preconditions discussed here include reduced access to forest land for the ethnic minorities and the poor, and the consequences of this in terms of reduced flexibility and reduced potential for reallocating resources for adaptation.

Adger et al. (2001) use the term 'social vulnerability' for the exposure of people to stress as a result of the impacts of environmental change. Social vulnerability encompasses disruption to livelihoods and loss of security. Social vulnerability may be a result not only of the environmental change in itself, but also of the policies addressing environmental change. In line with their argument this article looks at the inequitable consequences of policies addressing environmental risk.

Adger et al. (2001) find a relationship between response to immediate environmental stress and capacity to adapt to long-term environmental change, in that they are facilitated and constrained by the same architecture of entitlements (in the terminology of Sen, 1990). In essence the access to and distribution of resources affect household capacity both to cope with disasters and to adapt to long-term stress.

The approach to sustainable adaptation taken in this article acknowledges that households need diverse sources of income and food in order to spread risk and provide buffer incomes when the main source of income fails. It is argued that this is an increasingly important factor for households to be able to adapt to and live with increasing uncertainty and frequent losses due to climate-related stress.

2. Increased vulnerability and reduced access to forest land

2.1. Access to the forest land and coping with floods and climate stress

Fieldwork in A Luoi district after the 1999 floods suggests that the vulnerability of the ethnic minorities in the mountain areas has increased as a consequence of policies that move them away from forest-based livelihoods to reliance on intensive paddy rice production on river banks. Hong Ha commune was resettled in the Bo River valley of A Luoi district (Thua Thien Hue province) in 1976 as part of the government's ambitions to protect the forest and reduce swidden cultivation (Bao, 1999). The commune was allocated land on the river banks. The 1999 floods heavily eroded the river banks, settled large amounts of sand and stone on the paddy fields and inundated the cassava fields so that the roots rotted. The farmers lost most of their production and could not replant for months, because they had to manually dig away 0.5–2 m of stone.[1] Households survived through aid and by collecting leaves in the forest to sell. Access to forest resources played a crucial role in enabling people to cope and recover from the disaster (Table 1).

The shortage of land for food production became particularly acute after the 1999 floods. The household of Mrs K.M, Pa Rinh village, Hong Ha commune is one of many who lost their paddy land in the floods and had to depend on dryland crops and minor forest products for their living. She says:

> We used to have 1500 m^2 paddy rice field in the Khe Ca Te area. After the floods that land is now dry and sandy, but we can grow some dry land crops there. We have 500 m^2 of old swidden fields. This is now the most important source of food security for us. We would like to expand it, but we are not allowed to. We do not open up new fields, we return to old ones, which are left fallow, but they have very low productivity. We had 300 plants of cassava further away, but it is difficult to protect and the wild pigs destroyed it (Interview 17 September 2000).

According to the Hong Ha commune People's Committee, in 2000 people returned to old fallow fields and planted 40 ha of hill rice (lúa rẫy). People used a patchwork of opportunities

TABLE 1 The author's interview results from 45 households in Hong Ha commune, May–October 2000

Main source of food for coping	Number of households
Collecting minor forest products	6
Cultivating old swidden fields	3
Forest work (planting, tending)	6
Have state pension due to war injuries	6
Borrowed money from relatives	2
Government credit	3
Food relief	19

for food production, including the use of marginal areas, risk-prone areas and areas not meant for production at all. Mrs Xuong (interview 1 July 2000) in Con Tom planted paddy rice in her fishpond and dry-land rice on an old helicopter terrace on the hillside.

Collecting forest products was mentioned by several households as an important coping strategy, especially for the poor, but it was very hard work. On a return visit to Mrs Kan Mua's family in October 2000, only the children were at home. Both parents had been away for a month collecting rattan in forests far away.

2.2. Tension between forest protection and livelihood needs

Mr Nam from the commune People's Committee explains that most of the forest around Hong Ha commune is managed by the Bo River Watershed Management Board, which is a state board under the province Department of Agriculture and Rural Development. The village discussions in Pa Rinh (19 September 2000) suggest that as long as the villagers had an income from planting and tending they were basically positive about the arrangement with the board. However, since 2000, there have been no more forest-planting contracts.

The tension between risk reduction through forest cover and risk reduction through production development is also an institutional tension between different levels of government. Since the establishment of 'programme 327' in 1992, government policy has had a major focus on reforesting the hill slopes. Watershed management policies prioritized comprehensive forest cover, partly with the purpose of reducing the risk of dangerous flood levels on the low land. Forest management of major areas of A Luoi district was allocated to the Bo River Watershed Management Board under the province Department of Agriculture and Rural Development. Between 1994 and 2000 the board organized the planting of a major part of the hill land in Hong Ha commune, mainly with acacia. Agricultural production has gradually been transferred to the river valleys instead of the hill slopes, in order to protect the forest (Bao, 2002; Dupar and Badenoch, 2002). The immediate reaction of the government after the 1999 floods was to call for intensified tree planting. The Vietnamese newspaper *Lao Động* reports from a meeting for flood recovery on 15 November 1999 where the Prime Minister announced the implementation of the 'five million hectare forest programme' in the central provinces as a priority.

The perspectives of the mountain population were not taken into account. The settlements and production of Hong Ha commune are at greater risk to flash floods now that they are situated in the river valley. The lack of access to the hill slopes was perceived both by households and commune authorities as the main constraint for recovery from the 1999 floods, as well as the main problem in general.[2] According to district statistics the land for food production in Hong Ha decreased from 135 ha in 1996 to 104 ha in 1999 because of forest plantations.[3]

According to the households interviewed, the previous mode of production involved shifting cultivation to small plots in the forest that were not at risk to floods, as they were protected by surrounding trees. However, they do not argue for a return to this mode of cultivation. Households and commune authorities argue that erosion control and protection is workable with agro-forestry production on the slopes, and in that way combines environmental and livelihood needs.[4]

During the 1990s, the relations between the Hong Ha commune and the Watershed Management Board was still perceived as beneficial to the households, because they received tree-planting contracts.[5] Now that almost all commune land is planted with forest, staff of the commune People's Committee say that it is as if they have 'woken up'. They have started negotiations with province authorities and the Board to diversify land use, that is, to be allowed to harvest some of the trees to be able to intercrop with profitable species and food crops.[6] The A Luoi district People's Committee backs up the commune perspective.[7] The Vice Chairman argues that watershed forest could be allocated to households for protection as well as their economic benefit if it is close to their settlements. Even the head of the Forest Protection Station in A Luoi argues that agro-forestry in protected forest should be allowed, as people do not have enough food.[8] Mr Son from the district agriculture section is frustrated by the mismatch between forest protection and livelihood development strategies. He says:

> We develop the forest in order to reduce the impact of the floods. But there is too little research into how to plant and in which areas to plant in order to be effective. People will destroy the forest if they do not have enough land for agriculture production.[9]

In 2004, a research project at the Hue University of Agriculture and Forestry brought the province, district and commune levels of government together with the Bo River Board in a workshop, whereby the province representatives for the Department of Agriculture acknowledged the need for more integrated land use in Hong Ha commune.[10] The Board, however, is reluctant to agree to harvesting or allocation of the forest to the commune under their mandate (see Note 10). Contact with the project and Hong Ha commune in subsequent years revealed that the situation was still the same in 2009 in terms of allowing agro-forestry with food production. However, part of the forest has been harvested to allow for households to plant rubber trees.

2.3. Food security strategies are based on more intensive rice production

To achieve food security through the small areas of paddy land allocated to the households in Hong Ha commune, productivity has to be raised. The government is encouraging the use of high-yield rice varieties, which require high inputs of fertilizer and pesticides.[11] Although many farmers are happy with the higher yields, the investment costs remain an unsolved problem as the rice is for household consumption, not for sale.[12] So far rice yields have been low because people have been reluctant to invest a lot in fertilizer. Access to inputs from A Luoi town 23 km away is cumbersome because few households have means of transport. The commune is discussing plans to establish an input supply service at the commune level. Because people have very little cash, it would require a credit system, whereby the farmers can access inputs on credit and pay after harvest. However, this requires that the households have income from some other source, as the rice is only enough for consumption. Higher investments make people more vulnerable to seasonal crises and reduced harvest. The main attraction of the new varieties seems to be the shorter growing period, which reduces risk.[13]

2.4. Government emphasis on cash crop development

The government development strategy for the mountains emphasizes income generation through cash crops (People's Committee Thua Thien Hue, 2001, 2008). There has been a series of efforts by the state to develop profitable export-oriented crops. During the latter part of the 1990s, investments were made in cinnamon and sugarcane, which failed due to lack of market demand. From 2000 to 2002 the Department for Fixed Settlement encouraged production of pineapple and pepper, which has now also been more or less been dropped due to low market prices. During 2004–2005 all effort was put into encouraging rubber production.[14] According to the commune People's Committee, in June 2005 at least half of all households in Hong Ha were planting rubber trees. Inputs for planting are provided as credit in kind. Labour costs are paid for but add to the credit, which is to be paid back by the households as a percentage of the harvest every year, once the trees give harvest after 7–8 years. According to staff of the Hue University of Agriculture and Forestry, many people did not have a clear idea of the credit conditions, and some were not even aware that they were meant to pay back the money they received for their labour. It is not clear what happens to the credit if the project fails to give the expected returns. The commune party secretary[15] was critical of the government 'campaign crops' and stressed the importance of supporting households to develop different strategies depending on their different situations and conditions. The high prices on rubber at present, however, encourage many people to want to join the project.

In group discussions in Pa Rinh village,[16] the main line of argument to reduce vulnerability was to spread risk. Ways to do this include cultivation in different types of land with different risk pictures, and cultivation of many different crops. Traditionally, people have responded to high-production risks by not investing so much in any one area of production. The present development strategies require higher investments.

Schipper (2007) argues for the need to focus on reducing vulnerability in order to allow people to adapt. The case of Hong Ha commune suggests that the reduced access to forest land is making the mountain population increasingly vulnerable because they have no alternative source of income and food when floods, storms, drought and insect infestations cause production losses. Buffer resources to cope with such losses form a foundation for enabling people to get out of poverty and allow resources for adaptation.

2.5. The vulnerability of resettled households

Construction of dams and reservoirs is becoming more common and given high priority in development strategies in Thua Thien Hue Province, not only because they are a source of energy (hydropower) but also because of their role in flooding and salt water mitigation (People's Committee Thua Thien Hue, 2008). Dam and reservoir construction also involves the resettlement of households from the areas being flooded by the dams.

According to the Hong Ha Commune People's Committee chairman,[17] Pa Rinh village will have to move anyway, as the Province authorities took a decision in 2004 to build a hydropower dam on the Bo River, which means that Pa Rinh village will be permanently flooded. The commune leaders are worried about the long-term livelihood opportunities for the Pa Rinh population. Even though the compensation in money may be significant, the reduced access to land still leads to severe difficulties (see Note 17).

Ty (2009) has studied the resettlement of a Ka Tu (an ethnic minority) village in connection with the construction of the Binh Dien reservoir and hydroelectric plant in Huong Tra district, Thua Thien Hue. The study shows how people's livelihoods are undermined when they lose access to common property resources. It further suggests that people have too little land to make a living, despite government policy declarations that resettled households should enjoy improved living conditions after resettlement.

In some ways, the lives of resettled people became more comfortable. They had better houses and were closer to schools, clinics and roads. However, they lost land and access to forest and fishing waters, which means a decrease in both income and food security. The money they received as compensation (25 million VND per household on average) is frequently used to buy food, which is not sustainable. Their land for agriculture production was reduced from 77 to 10 ha. Before resettlement most of their income came from bamboo and rattan. Access to forests and rivers used to be an important source of rattan, honey, hunting and fishing. Now they have lost this access (Ty, 2009).

In recent years, the dam projects also undermined other state investments. The construction of the Ta Trach reservoir will flood large areas, which have been planted with rubber trees in the last five years, as well as put major investments in irrigation infrastructure underwater. Hong Ha commune in A Luoi district will lose a third of its rubber plantations and half of their irrigated paddy land, according to the Party Secretary of the commune (interviewed by the author, April 2009).

National and province perspectives on mountain area development are largely based on a model of intensive agriculture in a small area (Ministry of Agriculture, 2001). The mountain population are 'encouraged' to change their livelihood from forest-based production requiring large areas of land for crop rotation and fallow practices to follow a 'lowland model' of intensive rice cultivation close to the rivers. Practices of hunting, fishing and collection of minor forest products are also restricted and 'forgotten' when compensation for resettlement is decided.

2.6. Forest land allocation reduces access for the poor

The income gap between lowland and mountain households has increased over the past decade (World Bank, 2003). Swinkels and Turk (2006) argue that problems of access to land, especially forest land, may partially explain the fact that poverty reduction among ethnic minorities has been much slower than among the majority 'Kinh' population. Swinkles and Turk show that the allocation of forest land and perennial crop land has often been based on ability to invest in the land, which has resulted in the exclusion of many poor from the allocation of such land. They refer to 2004 statistics from the Ministry of Natural Resources and Environment showing that despite large dependency on forest land, only 24% of the ethnic minority households report having formal access to such land.

The process of forest land allocation in Thua Thien Hue province has been studied by Hoang Thi Sen (2009). Her research in two communes in Thua Thien Hue province found that poor households got less land allocated to them compared to better-off households. The average land holding in Loc Tien commune (lowland) was 8.25 ha for non-poor households and 1.45 ha for poor ones. In Thuong Quang commune (mountain) it was 4.75 and 0.7 ha, respectively, for non-poor and poor households. She draws attention to the experience of poor households, in particular poor women, who have reduced access to firewood and minor forest products as a result of the devolution of land rights. She also finds that the devolution of forest land has led to conflict between land-rights holders and households with grazing cattle, as well as conflict between households expanding land for forest planting and poor households who have been using the land for food security. Conflict has also arisen as households perceived the land allocation to be unequal, benefiting local leaders, while creating a loss of traditional access rights to the forest for the remaining households (Hoang Thi Sen, 2009).

2.7. Resettlement of lowland households to the mountains

Government efforts to adapt to floods, sea-level rise and salt water inundation include the resettlement of lowland households to the mountain areas, which further increases competition for land resources. As lowland households face increasing risk to their agriculture production, their interest in forest land increases. In this research, households were interviewed in Van Thanh village in the lowlands of Hai Lang district, Quang Tri Province, where 26 households (out of 230 in the village) were preparing to resettle in the mountains of Huong Hoa district. The provincial government arranged for land allocation to them as well as start-up capital. Other experience of such resettlement has shown that the allocated land is often used as common property by the local ethnic minorities (Chinh, 2001). The resettlement of lowland households in the mountains of Thua Thien Hue is not yet a major issue, but is expected to become one if production conditions in the lowlands become increasingly difficult due to climate change.

If allocation of land and forest resources is perceived to be inequitable, this is likely to lead to tensions, which constrain the negotiation process that is required to manage reallocation of resources. Adaptation cannot be sustainable without mechanisms for land allocation and resettlement that adequately compensate people who lose land, either as a consequence of climate impact or of measures to mitigate these impacts.

3. Conclusion: reduced adaptive capacity

The evidence presented in this article indicates a trend of increasing inequity and loss of access to forest resources, which previously have been vital to poor households in the area. It is argued that these losses reduce adaptive capacity in several ways. Firstly, the ethnic minority households concerned lose buffer resources, which previously helped them to cope with shocks, floods and drought that are becoming more and more frequent. Secondly, they are 'moved' into a development model of input-intensive farming, which increases the economic consequences caused by losses, that is, a crop loss means going into debt to input suppliers or banks. Thirdly, because the land area available to these minorities for agriculture is small, they have less flexibility and less opportunity to reduce risk through diversification. Fourthly, the community suffers a loss of adaptive capacity, which previously had opportunities to reallocate resources within the commune.

The development trends lead to fewer opportunities, both for individual households and for communities, to diversify and reallocate resources in response to changing conditions. Climate change increases levels of uncertainty, and it becomes increasingly risky to 'put all one's eggs in one basket'. At present the ethnic minority households in the mountains are being 'cornered' into putting all their resources into a single type of input-intensive investment. This counters the flexibility that must be a necessary attribute of any sustainable adaptation strategy.

Adger et al. (2002) are concerned that increasing inequality in access to resources may reduce communities' resilience by undermining social organization and reducing social cohesion. They caution that differential access to scarce environmental resources can undermine collective institutions, which exacerbates the vulnerability to external shocks of the marginalized.

Processes that increase resilience for some but not for others, and thereby increase inequity in society, cannot be considered sustainable. The inequity and difference between lowland and mountain communities and their interests and perspectives may lead to increased tensions in the future as competition over natural resources increases with climate change.

Policies that address climate impacts may also be undermining adaptive capacity among forest-dependent minority communities in the mountains. This research illustrates that adaptation of different groups of people has to be seen in the context of interaction between social and environmental processes. Social sustainability needs to be addressed in the context of land distribution, access to forest resources and policies for agricultural and forestry development, considered from the perspectives of livelihood risk and social cohesion.

Notes

1. Household interviews Hong Ha commune, May and October 2000.
2. Pa Rinh village group interview, 22 February 2001 and meeting with Hong Ha Commune People's Committee (CPC), 23 February 2000.
3. Meeting at the A Luoi Land management office, 28 September 2000.
4. Interviews and meeting CPC, 23 February 2001.
5. Interviews, October 2000.
6. Meeting Hong Ha CPC, 23 February 2001.
7. Mr Thoi A Luoi district People's Committee (DPC), June 2004.
8. Mr Chau, 21 February 2001.
9. Interview, 22 September 2000.
10. Hue University of Agriculture and Forestry (HUAF) staff, June 2004.
11. Staff of the Agriculture Section, March 2001.
12. Household interviews and CPC, June 2004.
13. Pa Rinh household group discussion, 22 February 2001.
14. A Luoi DPC and HUAF, June 2004.
15. Interview, 4 March 2001.
16. Pa Rinh household group discussion, 22 February 2001.
17. Meeting, 10 June 2004.

References

Adger, N., Kelly, M. and Ninh, N. H. 2001. *Living with Environmental Change. Social Vulnerability, Adaptation and Resilience in Vietnam.* Routledge, London, UK, and New York, NY, USA.

Adger, N., Kelly, M., Winkels, A., Luong, Q. H. and Locke, C., 2002. Migration, remittances, livelihood trajectories and social resilience. *AMBIO*, 31(4). 358–366.

Bao, L. Q., 1999. Management of forest resources in Hong Ha: existing situation, dynamics, and paths towards sustainability. *Community-based Natural Resource Management in Hong Ha*, L. V. An (ed.). Annual Report 1999, Hue University of Agriculture and Forestry.

Bao, L. Q., 2002. The system of land tenure and the forest tree and land use in Hong Ha. *Community-based Upland Natural Resource Management 1998–2001*, L. V. An (ed.). Agricultural Publishing House, Hanoi, Vietnam.

Carew-Reid, J., 2008. *Rapid Assessment of the Extent and Impact of Sea Level Rice in Vietnam.* International Centre for Environmental Management, Brisbane, Australia.

Chinh, N. V., 2001. Sedentarization and poverty reduction in Vietnam. *Paper for the Workshop on the National Poverty Reduction Strategy at MOLISA*, Hanoi, Vietnam.

Dupar, M. and Badenoch, N. (eds), 2002. *Environment, Livelihoods and Local Institutions: Decentralization in Mainland Southeast Asia.* World Resources Institute, Washington, DC.

Hanh, T. T. H., 2010. Income diversification in fishing and aquaculture in the Tam Giang Lagoon – adaptation to climate change or not? MSc thesis. Hue University of Agriculture and Forestry and SLU, Sweden. www.sol.slu.se.

IPCC (Intergovernmental Panel on Climate Change), 2007. *Fourth Assessment Report: Climate Change 2007 – Synthesis Report.* www.ipcc.ch/pdf/assessment-report/ar4/syr/ar4_syr.pdf.

Ministry of Agriculture, 2001. *National Five Million Hectare Reforestation Program (1998–2010).* Hanoi, Vietnam.

Nelson, D., Adger, N. and Brown, K., 2007. Adaptation to environmental change: contributions of a resilience framework. *Annual Review of Environment Resources*, 32. 395–419.

People's Committee Thua Thien Hue Province, 2001. *Report on the Socio-Economic Development Plan of Thua Thien Hue Province*. www.thuathienhue.gov.vn.

People's Committee Thua Thien Hue Province, 2008. *Reports on the Construction of the Binh Dien Hydropower Dam*. www.thuathienhue.gov.vn.

Phuong, L. T. H., 2010, *Climate Change and Farmers' Adaptation*. A case study of mixed – farming systems in Quang Tri province, Vietnam. MSc thesis. Hue University Agriculture and Forestry and SLU Sweden. www.sol.slu.se.

Schipper, L., 2007. *Climate Change Adaptation and Development: Exploring the Linkages*. Working Paper 107. Tyndall Centre for Climate Change Research, Norwich, UK.

Sen, A. K., 1990. Food economics and entitlements. *The Political Economy of Hunger*, Vol. 1, J. Drèse and A. Sen (eds). Clarendon Press, Oxford, UK. 34–52.

Sen, H. T., 2009. *Gains and Losses. Devolution of Forest Land and Natural Forest. A Study of Forest Allocation in North Central Coast, Vietnam*. Swedish University of Agriculture Sciences (SLU). www.epsilon.slu.se.

Swinkels, R. and Turk, C., 2006. *Explaining Ethnic Minority Poverty in Vietnam: A Summary of Recent Trends and Current Challenges*. Unpublished paper. Hanoi, Vietnam. www.worldbank.org/vn.

Thomalla, F., Downing, T., Spanger-Siegfried, E., Guoyi Han, and Rockström, J., 2006. Reducing hazard vulnerability: towards a common approach between disaster risk reduction and climate adaptation. *Disasters*, 30(1). 39–48

Ty, P. H., 2009. *Impoverishment Risks Due to Involuntary Displacement and Resettlement in a Hydroelectric Development Project*. RDViet Research Report. Hue University of Agriculture and Forestry, Hue, Vietnam. www.rdviet.net.

World Bank (coordinator), 2003. *Vietnam Development Report 2004. Poverty*. Joint donor report to the Vietnam Consultative Group Meeting. Hanoi, Vietnam.

research article

climate and development

Sustainable adaptation and human security: Interactions between pastoral and agropastoral groups in dryland Kenya

BERNARD OWUOR[1,*], WYCLIFFE MAUTA[2] and SIRI ERIKSEN[3]

[1]Kenya Forestry Research Institute, P.O. Box 1078-80200, Malindi, Kenya
[2]Kenya Forestry Research Institute, P.O. Box 20412-00200 City Square, Nairobi, Kenya
[3]Department of International Environment and Development Studies – Noragric, Norwegian University of Life Sciences, P.O. Box 5003, NO-1432 Aas, Norway

This article investigates how pastoral and agropastoral populations interact in adapting to climate variability and change, particularly to drought. Interactions within trade, livestock and human mobility, and accessing forest resources are critical to local adaptive capacity in Kenya's drylands. Qualitative interview data collected between 2004 and 2007 in Endau, eastern Kenya, are analysed to explore the role of these interactions in sustainable adaptation, and how they have been affected by formal policies and informal governance. The article also explores how politics, decision making and conflicts interact in practice to shape decision making, and how dominant state orientation may facilitate or constrain sustainable adaptation. We conclude that both official policy and state practice in terms of actual decision making (whether in line with policy and legal frameworks or not) appear to undermine human security in terms of political and social rights, as well as sustainable adaptation in terms of social equity and environmental integrity. Sustainable adaptation for the case of Endau would imply a fundamental change in governance regime from one of imposing punitive measures to stop dynamic interactions to one through which, instead, interactions between the various groups are strengthened.

Keywords: adaptation; climate change; conflict; drylands; Kenya; pastoralism; sustainability

1. Introduction

In this article, we investigate the role of flexible interactions among pastoral and agropastoral populations in sustainable adaptation to climate variability and change, and how these strategies may be supported or impeded by formal policies and state practices. Climate change is likely to have a great impact on dryland populations in East Africa. Climatic uncertainty will probably increase in terms of incidences of both drought and flood as well as shifts in seasonality (IPCC, 2007). Several political, economic, social and environmental constraints have also contributed to the vulnerability of dryland populations in sub-Saharan Africa (Twyman et al., 2004; Owuor et al., 2005; Ziervogel et al., 2006; Tschakert, 2007). At the same time, it is well documented that both pastoral and agropastoral populations have a rich set of strategies with which they continuously adapt to environmental and social change (Scoones, 1995; Mortimore and Adams, 2001; Reid and Vogel, 2006; Thomas et al., 2007).

Interactions between groups, in terms of mobility, trade and access to natural resources, have important implications for social equity as well as being linked to environmental integrity, that is, the key aspects of sustainable adaptation (Eriksen et al., 2011). Literature reveals divergent evidence regarding potential social and environmental effects of such strategies and interaction between populations. Adaptation can be seen as

■ *Corresponding author. *E-mail*: **benodit912@gmail.com**

ISSN: 1756-5529 (print)

autonomous or planned (Smit et al., 2000); that is, people's strategies (including practices and social relations) for managing variability and change form part of adaptation, in addition to deliberate actions by institutions to alter practices, processes, legislation, regulations and systems in order to reduce society's vulnerability (Burton et al., 2002; Eriksen and Lind, 2009). Eriksen et al. (2011) observe that although it is often assumed that all adaptation is benign for all groups, any response, whether local and autonomous or policy driven, may benefit some groups but at the same time have negative effects on other groups and the environment in the short or long term.

One school of thought in academic and policy discourse argues that climate change will lead to conflict over scarce resources, social strife and weakening of the state (Schubert et al., 2008). Following this interpretation, local strategies to adapt to climate variability and change, and interaction between groups in particular, can lead to social strife, and widespread negative effects on human well-being as well as overexploitation of natural resources. Implicitly then, local strategies are unsustainable and outside interventions are key to adapting to climate change. Such propositions are countered by studies showing the strength of household and community strategies and the need for any formal adaptation actions to build on such local vulnerability contexts (Mortimore and Adams, 2001; Tschakert, 2007). Interaction is a key ingredient of living with environmental change, and locally developed enabling frameworks for negotiation and cooperation are key to adaptation. According to this view, it is often state development efforts and policies that have created conflict, reduced local control over resources and access to them, and hence increased vulnerability and social inequity (Thébaud and Batterbury, 2001; Benjaminsen, 2008). Even if essential to adaptation in drylands, interactions between groups are not unproblematic or uncontentious, however, and are likely to have highly differentiated outcomes because material benefits, control over resources and power relations are all negotiated as part of political processes. This article investigates social equity and environmental integrity outcomes in terms of options and capacity for maintaining human, social and political rights more widely, that is, for achieving human security (O'Brien, 2006) rather than a focus on only human survival, material incomes and conservation of biodiversity.

The issue of pastoral–agropastoral interaction exemplifies how sustainable adaptation, in terms of adjustments that contribute to social equity and environmental integrity, is likely to be intertwined with complex social processes, such as natural resource access, exclusion, government policy processes, and economic and political structures and relations. These features make the interaction between governance (i.e. how formal and informal arrangements coordinate social behaviour) and sustainable adaptation particularly important. Governance describes the type and character of interactions between people, collective decisions and the role of institutions in negotiation processes and how local actors exercise power (Kemp et al., 2005). Formal and informal structures often interact in governance processes (Selboe, 2008). For example, in addition to developing formal, explicit policies, government institutions play an important implicit role in governance through political allocations, and by favouring particular groups or interests in individual decision making by, for example, imposing penalties and granting rights, and in resolving disputes (Kemp et al., 2005). Revealing dominant state orientations is therefore necessary for understanding sustainable adaptation.

This article investigates how policies, politics and practical decision making have hindered or helped interaction between pastoral and agropastoral populations in the drylands of Kenya. The study focuses on one hill, Endau, and its immediate surroundings, where Kamba, Orma and Kenyan Somali pastoral and agropastoral populations access both pasture and water during times of drought. Hilltop areas provide goods and services that are critical to production systems, cultural identity and spiritual values in the drylands, including the study area, Kitui District (Gachathi, 1996; Eriksen et al., 2006). Forest and water resources connected to the hilltop ecosystem play a vital role in supporting local adaptation mechanisms in these areas (Owuor et al., 2005; Eriksen et al., 2006). First, this article describes local adaptation strategies and interactions. It then moves on to analysis of qualitative data to distinguish how changing policies, government interventions, elections and local conflicts affect how pastoral and agropastoral groups interact and utilize dryland resources. There is a particular focus on how access to adaptation strategies by different groups has been negotiated through local institutions and resource rights.

There is a need to understand how adaptation responses interact with other processes of social and environmental change in order to determine how such responses can contribute to sustainability. So far, few adaptation policies have been developed or implemented in the area, and government policy and action to date may be more important where it interacts with local adaptation strategies, such as through regulating resource access regimes or prioritizing particular economic or political interests. Critically, this study shows that flexible interactions are essential components of sustainable adaptation but that these have been impeded by past and present policies and governance in the drylands.

2. Methodology

2.1. Study site

The study was conducted in the area around the Endau Hilltop Forest, which is in Kitui District in Kenya's Eastern Province[1] (see Figure 1). The area lies between the central highlands of Kenya and the coastal forests. The altitude ranges from 500 m above sea level (masl) in the dry lowlands to 1,400 masl at the highest peak. Natural resource use and farming are summarized in Table 1, and population and poverty statistics are given in Table 2. This area was selected because of the interaction between Kamba, Orma and Somali pastoral and agropastoral groups, especially during drought, the importance of the hilltop and related resources in these interactions, and a history of government intervention affecting local adaptation strategies.

2.2. Data collection

Interviews were carried out in four settlements around the hilltop: Ndetani, Ikisaya, Malalani and Twambui (see Figure 2). Qualitative approaches were used in data collection, including group interviews, key informant interviews and informal observations as well as household structured questionnaires. Sampling of households was conducted on the different sides of the Endau Hilltop Forest with the help of members of the local administration and local research assistants who had knowledge of the social and ecological aspects of the study area. Selection of respondents was done against a set of criteria including gender, distance from the hill, tribe, age and social status. During the study period, between 2004 and 2007, 132 household and key informant interviews (including farmers, pastoralists, local administration, leaders and craft sellers), 15 group interviews and eight market visits were carried out in Endau. In addition, five interviews were carried out with key informants in the Kitui District administration, and seven interviews were carried with agropastoralists and pastoralists in Mutha (a settlement further south) because some of the pastoralists who visit Endau have settled in this area.

The data collection was carried out during both dry and wet seasons, and the study area and informants were followed over several years in order to capture the seasonal variations as well as political processes, conflict and changes in power relations over time (the dry seasons of August 2004, February 2005, September 2006, February 2007 and October 2007, and the wet seasons of December 2005 and November 2006; the February 2005 and October 2007 fieldwork periods also covered the run-up to a national referendum over

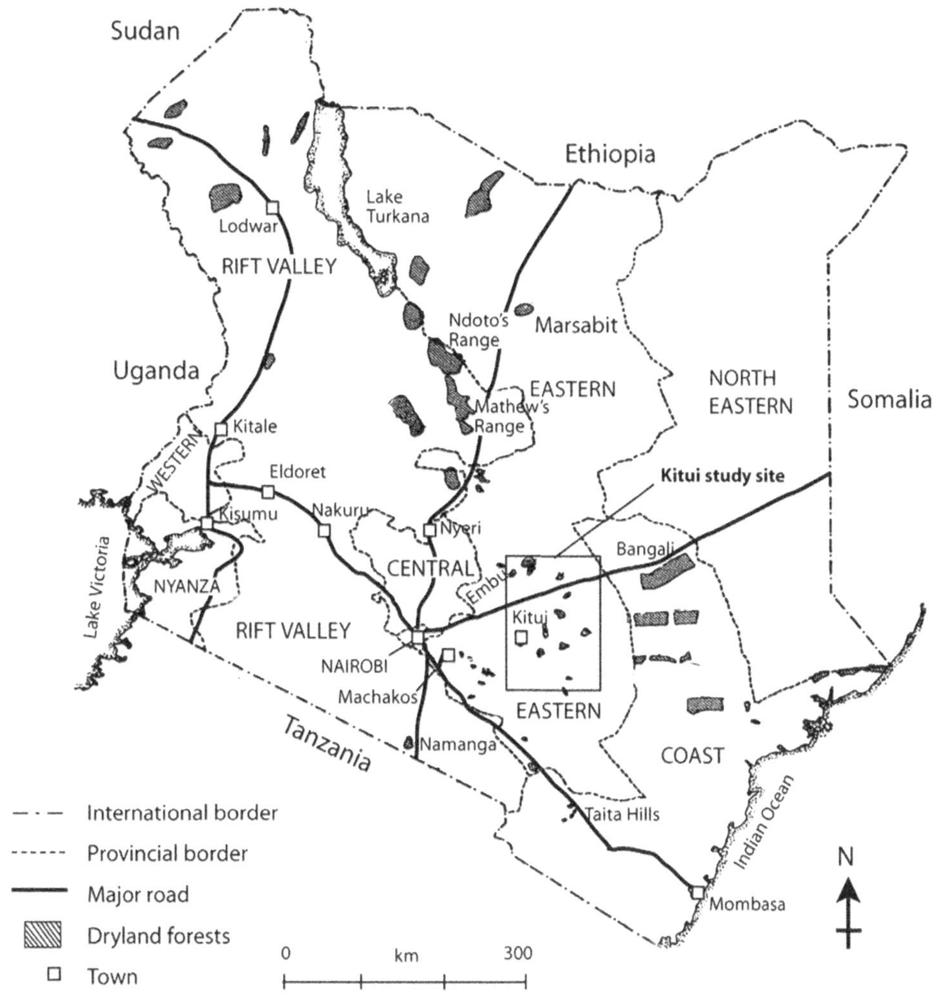

FIGURE 1 Location of Kitui study site in Kenya

constitutional reform and the December 2007 general elections, respectively). Figure 3 shows the seasonal variations in rainfall distribution. Interviews covered topics such as human settlement and seasonal movements, marketing and trade, forest uses, access and management of natural resources, politics, negotiation of interests, conflicts, development interventions and conservation policies. A specific ethnobotanical survey (including five transects and five farmer visits) was also conducted to find out how pastoral and agropastoral people access and utilize the area's natural resources. In addition, archival studies were conducted by visiting the national archives to collect historical data to complement the primary information and for historical analysis of the study area and communities.

3. Analysis and discussion

3.1. Local adaptation strategies and interactions

Endau becomes a melting pot of cultures and lifestyles during the dry season from July to October

TABLE 1 Environment and natural resource use in the Endau case study area

Climatic conditions	Water sources	Agriculture	Grazing	Forest uses
780 mm annual rainfall[a] Two rainy seasons, October–December rains most important Failure of rains 1 in every 2–3 years Long dry season Intermittent floods	Seasonal springs Seasonal rivers Shallow wells on northern and eastern sides of hill Piped water on southern and western sides of hill, sourced from the hilltop Boreholes	Livestock herding important on north and east sides of hill Rain-fed agriculture on the southern side of the hill and along rivers Historically cultivation on the hill	Seasonal pattern of wet-season grazing on plains and dry-season grazing near or on the hill Also grazing on private land, for rent	Historical use of indigenous forest on Endau hill Presently indigenous plant products mainly accessed from lowland forests for home consumption and small-scale marketing

[a]Estimate based on rainfall data for 0.5° × 0.5° grid covering eastern parts of Kitui District (mean annual rainfall in 1961–1990, 779.9 mm) as well as data from the nearest rainfall station, Zombe (mean annual rainfall in 1961–1990, 776.6 mm).
Source: Adapted from Eriksen and Lind (2009).

TABLE 2 Population and poverty statistics of the case study area

District/location/ sub-location	Number of individuals	Percent of individuals below poverty line
Kitui District	489,747	70.48
Ndetani sub-location	2,546	63.30[a]
Syou sub-location	1,264	58.10[a]
Malalani sub-location	1,438	58.10[a]
Twambui sub-location	1,665	58.10[a]

[a]Estimated from location-level poverty data.
Adapted from Eriksen and Lind (2009).
Source: Government of Kenya (1999), Republic of Kenya (2003), and the Kitui District Information and Documentation Center population statistics as of 2003.

as visiting pastoralists interact with sedentary agropastoralists near the hill. The interaction intensifies during prolonged drought, such as when the long rains failed in 2005 (described by household and group interviews as well as informal observations during December 2005 data collection). The interaction involves several ethnic groups. According to key informant interviews with elders in August 2004 and February 2005, Kamba groups originally started visiting Endau 150 years ago in search of pastures for their cattle, displacing other pastoralist groups such as the Akavi. The Kambas gradually settled to become agropastoralists, farming nearby and on the hill and migrating with livestock far onto the plains in the wet season. Some Kambas moved livestock onto the hill during severe droughts. Kamba agropastoralists also settled in the neighbouring district, Mwingi. According to key informant interviews with resident Kambas during several of the data collection visits as well as with Mwingi pastoralists in Twambui in September 2006, Mwingi Kambas have recently begun to send their cattle, mainly with hired herders, to Endau for water and grazing during the dry season. Orma and Somali pastoralists also migrate with cattle to Endau during the dry season, sometimes herded by the owners themselves and at other times by hired herders. Interviews with Orma and Somali herders in November 2006 revealed that some of these

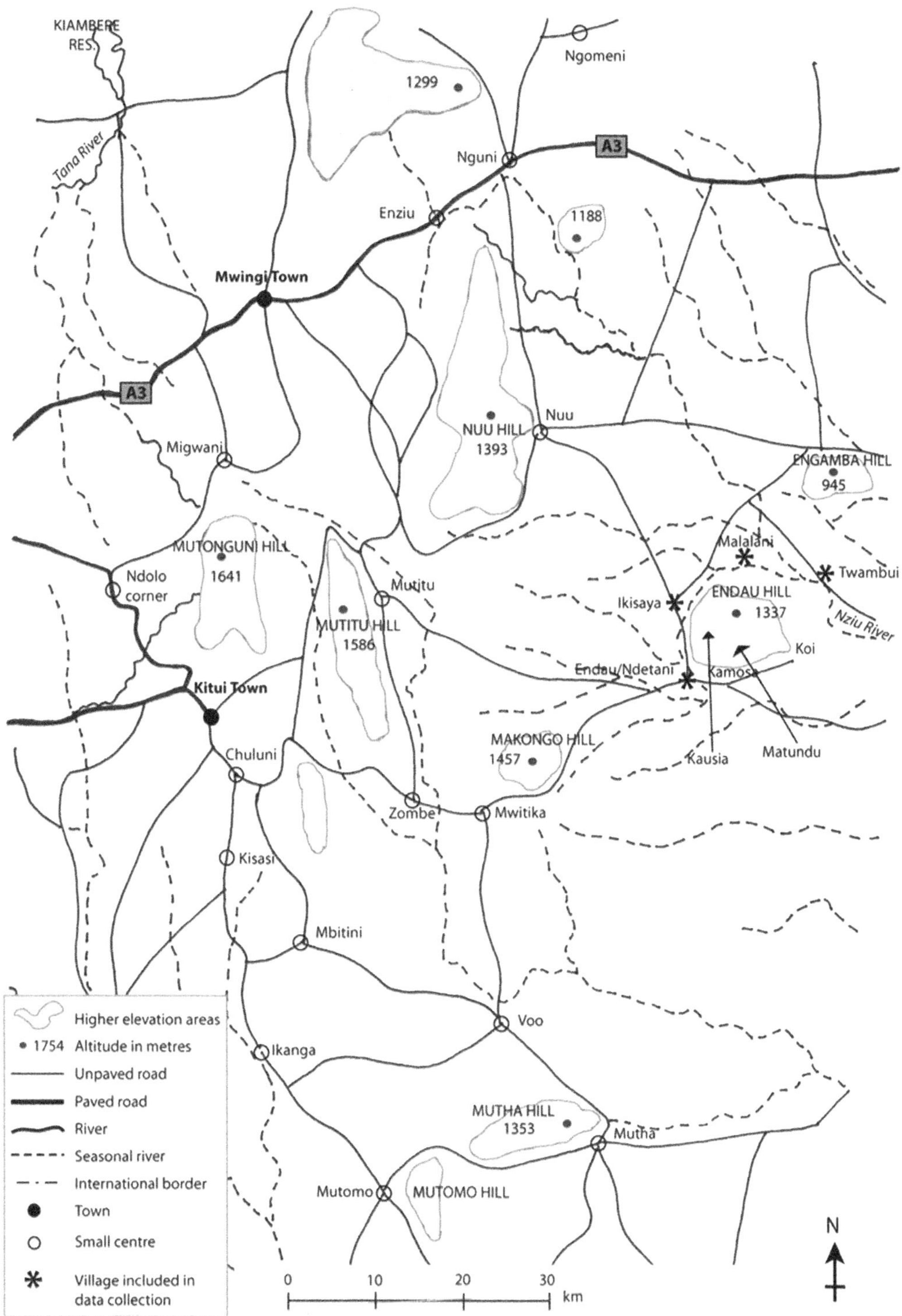

FIGURE 2 Location of settlements (Ndetani, Ikisaya, Malalani, Twambui and Mutha) where data were collected

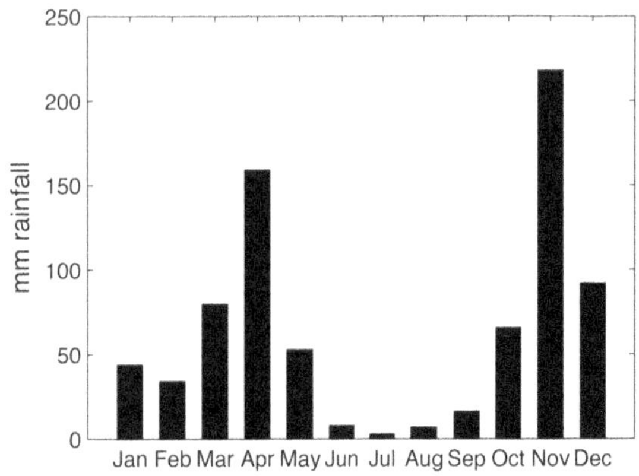

FIGURE 3 Mean monthly rainfall 1961–1990, Kitui. Based on rainfall data for 0.5° × 0.5° grid covering eastern Kitui District, CRU TS2.1 dataset

cattle owners live in permanent settlements and farmland in districts north and east of Kitui, and others are going through a transition to agropastoralism through settling in Mutha, south of Endau. These movements are taking place together with numerous other adjustments; for example, Somali and Orma elders indicated in our interviews in 2006 that when a drought approaches, they reduce their stock in general and the number of animals most vulnerable to drought in particular, such as goats and sheep, and keep more camels, which can survive prolonged periods of drought.

Various strategies are adopted to cope with drought. The resident agropastoralists rent out wells and pastures to the visiting groups, gather wild fruit and food, and hunt wildlife for meat. Trade with the pastoral groups increases, particularly the sale of cereals, and the frequency and quality of meals tend to reduce. The pastoral groups travel long distances to areas with better access to water and pasture, and loan their livestock to relatives in different areas to reduce risk, thus reducing their stock and altering herd composition. Forest surveys and ethnobotanical household interviews in 2005 and November 2006 revealed that 20 streams emanate from the hilltop forest. Resident Kambas in Ikisaya, Ndetani and, to some extent, Malalani use a number of products from the forest on the hill and nearby plains. Wood and timber are used for house construction, firewood and charcoal, wood carving, and making tools and other implements. Plants are used for traditional medicine, and wild fruit and honey are collected for food. Orma and Somali groups, as well as Kambas from Mwingi and those settled in Twambui (who are further from the hill and more oriented towards pastoralism) generally depend on forest products from the plains rather than from the hill. As described in more detail by Owuor et al. (2005), Lind and Eriksen (2006) and Eriksen and Lind (2009), the hill is critical for local survival during drought through alternative grazing sites for cattle, wild food, sale of timber and charcoal, hunting and traditional drought ceremonies. The hill, with its moister and cooler microclimate, is also a watershed for rivers, thus providing crucial services such as dry-season wells and water for crop cultivation at the foot of the hill. Interaction through renting of the wells and trade in food crops is central to Orma and Somali drought strategies.

However, interaction is not without problems. As explained by Lind and Eriksen (2006), a period of violent raiding and attacks on Kamba settlements by Somali groups occurred from the 1970s until the 1990s, but since that time more peaceful interaction and trade have increased. Renting out to Kamba, Orma and Somali pastoralists of dry-season wells dug in the river bed (as well as grazing land) is normally a major source of income for the local Kamba agropastoralists. The trade in livestock with Somalis and Ormas is also an important means of improving local livestock breeds. The agropastoralist Kamba in Endau mostly sell cereals to generate income and to buy livestock. This trade has been one of the main sources of drought livelihoods for Kamba agropastoralists in Endau over the past few years (Eriksen and Lind, 2009). The interaction has become particularly important to settled agropastoralists over the past decade after people were evicted from the hilltop in 1996 as part of government conservation priorities, and they lost access to highland farming, grazing and forest products on which they had relied in their strategies to manage drought. These findings concur with observations in other dryland studies, that is, pastoralists depend on movement of livestock and people seasonally to look for pasture and water as a key way of managing climatic variability and uncertainty (Scoones, 1995; Oba, 2001). The current study shows that local strategies and interactions are important for settled agropastoralists as well. Pastoral–agropastoral interaction has become critical for survival and to cope with drought for all groups in the area. In addition, the mutual dependency and cooperation that it represents plays a crucial role in maintaining peaceful relations.

Hence, three main types of interactions are important in managing seasonal variations in rainfall and drought: the movement of cattle to the plains during the rains and nearer to the hill during the dry season; trade between pastoral and agropastoral groups during drought; and the use by Kamba agropastoralists of forest products and services provided by the hill. The findings underscore the need to build on local strategies and interactions as an essential component in sustainable adaptation, rather than imposing highly centralized adaptation measures. The interactions contribute to the human security of pastoral and agropastoral groups in general, reducing overall poverty. Trade has allowed some impoverished people to gain new sources of livelihoods, possibly counteracting social inequity. The growing importance of trade has contributed to distributing economic and political influence to more actors within the community (Eriksen and Lind, 2009), which could be seen as increased political and economic equity. At the same time, Eriksen and Lind describe how the effects are uneven within groups and subject to intense negotiations. For example, material benefits from renting out wells disproportionately accrue to some families – namely the well owners – while threatening water access for others. In order to understand how interactions form part of sustainable adaptation, it is critical then to understand how politics, decision making and conflicts interact in shaping internal negotiations and their outcomes.

3.2. The effect of formal policy on interaction and sources of managing drought

The findings above suggest that, in order to ensure that economic activities are not disrupted and that people can survive prolonged drought, flexible policies are required. However, several formal policies have contributed to impeding interaction and adaptation strategies in the drylands. A key feature of government policy in the drylands since colonial times has been efforts to sedentarize pastoralists (Heald, 1999; Lind and Eriksen, 2006). Korir (1991) describes Kamba protests in 1937 against the government's use of force to destock and confiscate livestock. Furthermore, during the colonial period, land policies divided the country into 'White Highlands', in which land was cleared and mixed farming, plantation crops and commercial livestock husbandry were introduced, and the 'African Reserves', in which growing African populations were confined to cultivation for food and income. Here, colonial policies focused on soil erosion control, terracing, protection of river banks and other environmental protection measures, in effect controlling farmer practices. These policies excluded pastoralists from many of their traditional grazing areas and restricted livestock movement. The colonial laws also excluded pastoral populations from the dairy and livestock markets. As a result, populations were disenfranchised and impoverished.

McSherry and Brass (2007) argue that many of the disfunctionalities and anti-poor biases of the livestock sector created during colonialism are still present today. Korir (1991) describes how the period just after independence had a policy regime for the drylands where soil conservation regulations remained in force but with weak enforcement. Post-independence policies then increasingly focused on experimentation with various development plans for the drylands. These included policies and enactment laws emphasizing environmental conservation and improvement of livelihood, for example, through agricultural education, training and extension work. Hence, focus has remained on protecting the environment, controlling local resource use and introducing new, more 'modern', techniques rather than facilitating existing livelihood activities such as pastoralism. At the same time, there have been few efforts to update and strengthen policies to facilitate livestock production despite great changes in economic framework conditions (e.g. through liberalization). For example, the Dairy Bill has not been updated since the 1950s, and the Livestock Development Policy and Dairy Development Policy have not been modified since the early 1980s and 1990s, respectively. The 2004 draft policy for sustainable development of the arid and semi-arid lands has not yet been passed, meaning that there has been no new policy for drylands development since the early 1980s (McSherry and Brass, 2007). The efforts that have taken place have been ineffective in countering marginalization of dryland populations. For example, a pastoralist parliamentary group was created in the early 1990s, concerned with land rights, droughts, food aid, livestock marketing and insecurity issues, but has had very weak impact according to analysis by Morton (2005). However, the recent attempts to redraft the Kenyan Constitution (passed 2010) have led to growing calls to reform current land laws, a move that would, according to McSherry and Brass, be crucial to pastoral livelihoods. They also see increasing understanding among some policy makers that pastoralism can be viable, evident, for example, in the recent enlargement of the government's Arid Lands Resource Management Project.

Interviews with Kamba key informants confirm that sedentarization has been a long historical process, recently accentuated by violence and raids that led to livestock losses and a search for safer livelihoods near the hill. Group discussions with Orma and Somali pastoralist groups confirmed that they too are undergoing sedentarization processes, while Mwingi Kambas are increasingly forced to drive their cattle all the way to Endau because of government land-use policies that involve privatization and cultivation of dry-season grazing areas.

In addition to the policies above, forest conservation policies have had a particularly negative effect on interactions and adaptation strategies in Endau. Conservation policies enacted through the Forest Act, Chapter 385, have involved an increase in the number of protected areas in the drylands such as Kitui District (Eriksen et al., 2006). This development has excluded pastoralists and agropastoralists from important sources of water, forest products and drought grazing areas and limited the area available for cultivation. Table 3 points to the fact that such exclusion, which has been experienced repeatedly in Endau, has been important in creating conflicts that further impede flexible interactions. The Endau Hilltop Forest is a gazetted forest managed by the government as a catchment area and ecologically sensitive ecosystem (Owuor et al., 2005). Since the most recent eviction of people from the hilltop in 1996, a permit has been required to enter the forest.

However, over the years, concerns have been expressed over existing laws and legislation, in particular that they fail to consider local needs and do not effectively advance local forest use. This has led to pressure from local communities, civil societies, donors and the private sector, resulting in the enactment of the Forest Act of 2005. This is aimed at sustainable management, including conservation and rational utilization of forest resources for the socio-economic development of the country (Government of Kenya, 2005). The Act acknowledges the role forests play in rural economies and makes provisions for how to involve those who live in and around forests in management and conservation of them.

However, though operational since 2007, the new Forest Act has so far not been able to forge the envisaged partnerships in the management of the forests. Interviews in Endau, with both the forest authorities and Kamba agropastoralists, revealed that the new bill has so far not been explained well enough nor put into practice in terms of being able to meet the needs and aspirations of the local populations. As a result, few benefits accrue from the hill to the adjacent population through formal mechanisms or measures.

TABLE 3 Conflicts and their impacts in Endau sites

Village	Types of conflict and insecurity	Proximate effects
Ndetani	Eviction of farmers from hill by the government	■ Landlessness and in-migration ■ Restricted dry-season grazing ■ Loss of fruit trees and drought crops
	Violent raids of Endau Kamba by Somali groups in the 1990s	■ In-migration of people exposed to violent raids in Twambui and Malalani
Ikisaya	Violent raids of Endau Kamba by Somali groups in the 1990s	■ Loss of livestock and lives ■ Closure of school ■ Disrupted farming ■ In-migration of people exposed to violent raids in Twambui and Malalani ■ Restricted dry-season grazing
	Conflict between villages over water pipeline from the hill	■ Sabotage to pipes and disrupted maintenance
Malalani	Violent raids of Endau Kamba by Somali groups in the 1990s	■ Loss of livestock and lives ■ Out-migration ■ Restricted wet-season grazing
	Eviction of farmers from hill by the government	■ Landlessness and out-migration ■ Restricted dry-season grazing ■ Loss of fruit trees and drought crops
	Interaction between pastoralist and agropastoralists over access to shallow wells and dry-season grazing	■ Precarious access for pastoralist to wells ■ High fees to rent water, free access to water difficult for settled farmers ■ Livestock theft ■ Local political tension in the village over increased trade and incomes during drought
Twambui	Violent raids of Endau Kamba by Somali groups in the 1990s	■ Loss of livestock and lives ■ Out-migration ■ Restricted wet-season grazing
	Interaction between pastoralist and agropastoralists over access to shallow wells and dry-season grazing	■ Precarious access for pastoralist to wells ■ High fees to rent water, free access to water difficult for settled farmers ■ Livestock theft and killing ■ Local political tension in the village over increased trade and incomes during drought

Source: Adapted from Eriksen and Lind (2009).

Although the hill has the only permanent water spring in the area, our visits during the 2004–2007 period revealed that only in Ikisaya is the water from the hill tapped and piped to the local population. Other laws still remain inflexible: for example, the Wildlife Act outlaws hunting, although this is a major coping strategy during drought. Market surveys revealed that in addition to being an important source of food, game meat is frequently sold during drought.

Although officially dismantled, many customary rules regarding forest access and use are still followed, meaning that flexible interactions take place informally. This informality means that the strategies are precarious and costly – people were regularly brought to court and fined. Conflicts ensued between the forest service officers who were mandated to take care of the forest and the local population taking their animals up the hill during drought.

Lind and Eriksen (2006) describe how Kambas became more vulnerable to drought after the 1996 evictions. Visits to the study sites at the peak of the dry seasons from 2004 to 2007, and discussions with staff of humanitarian and relief agencies including the Catholic Church, the drought management project under the office of the President, Adventist Relief Agency and the provincial administration, revealed that the people of Endau were net importers of food and had to rely on relief food supplies to see them through the dry seasons. The rapid response and high level of preparedness needed to cushion the local people were still lacking, despite drought being periodic and predictable. In 2007, an intense prolonged drought was declared a national disaster. People were not able to meet their food needs and had to rely heavily on relief food supplies to see them through the bad times, and yet relief supplies have seldom been adequate or timely.

Hence, a range of policies in combination with government neglect of local needs has made it increasingly difficult to keep livestock and to adapt herding to climatic variations. Associated destitution and loss of human security among several families interviewed in Endau suggest that social equity and sustainable adaptation are undermined by such policy outcomes.

3.3. The effects of government development interventions and political leadership on adaptation

Past government policies have marginalized people in drylands, contributing to geographical social inequities, not only by restricting people's livelihoods as described above but also by poor provision of infrastructure. The drylands have been seen as having low economic potential and being politically unimportant. As a result, government allocations for development in the drylands have often been low. However, observations in the case study area indicate that recent political changes at the national level, involving democratization and decentralization of development funding through creation of the constituency development fund and the local authorities transfer fund, may provide more investments and influence. There are a range of new and ongoing projects in the area. Dispensaries and schools are being built, and if completed and equipped with the necessary facilities to provide the intended services, the quality of life will be greatly improved. Roads are under construction (though, despite improved maintenance, roads remain impassable during the rainy season), and water projects are under way. The elected leaders have had leeway as to where to locate these projects. Several of the Kamba farmers expressed in interviews that these recent developments made them feel that their interests and needs were increasingly being considered, which may have contributed to local support for the current government during the latest elections in 2007.

Although these developments may have alleviated some historical–geographical inequities, the political nature of allocation of development funds locally also contributes to new social inequities. In a group interview with community leaders in 2006, respondents reported a need to involve local people in decision making on which projects are funded and which people or companies get the contracts to supply construction materials. To achieve this, more transparent and accountable ways are needed of vetting projects funded by the constituency development funds, which are managed by the members of parliament (MPs), and the local authority transfer fund managed by the councillors (local district politicians). The leaders also indicated that processes should be more participatory and inclusive and that the benefits of projects should be equitably distributed. For example, water is a major limiting factor to production in the area. The last five years have seen the sinking of two boreholes in the area, but their location is highly contested as they are located on private as opposed to communal land and give influential people easier access to water. Informants claimed that politicians favoured areas where they received more votes and awarded contracts to supply materials for various projects to themselves and close relatives. The elected leaders are also not equally accessible to all of their electorate, and some of the areas are very infrequently visited by politicians, if at all.

Both councillors and local MPs wield considerable power through allocating development funding, and also through local political campaigning: because of a high illiteracy rate, the local people have had to rely on local leaders' interpretation of political manifestos in public meetings. According to present and past leaders interviewed in this study, including MPs and councillors (both elected) as well as chiefs (administrators appointed by the government) and village elders (appointed by the chief in collaboration with the local population), much has changed in the political landscape with the introduction of multiparty politics in the 1990s. Elections are held every five years for both parliamentary and civic positions. During political campaigns, individual candidates and political parties go to the grassroots to get votes, using every opportunity to sell their political agenda, including at funerals, public meetings and local markets. They also spell out the plans they have for their constituencies at functions organized by the provincial administration. Development funding (and promised investments in schools, dispensaries and other infrastructure) is therefore used to mobilize political support and strengthen the power of local politicians, which contributes to social inequities felt locally.

3.4. Informal governance, the politics-policy nexus and sustainable adaptation

The issue of movement of people during drought illustrates how national and local politics influence decisions that may not be part of official policy or legal frameworks but that directly undermine local adaptive capacity. Restricting the movement of pastoral communities, which was a deliberate colonial policy, continues through actions by district administrations and local politicians in particular. During the colonial period, Kenya was divided into districts and assigned to ethnic groups. The authorities controlled movement by making it illegal to cross district boundaries without a special pass. This is confirmed by our archival studies, which revealed communication between the colonial chiefs and the District Commissioners of Kitui and the neighbouring Tana River District. The Kenyan constitution now guarantees freedom of movement and no formal laws restrict mobility. In practice, however, such movement between different areas of the country is still severely restricted along ethnic lines through decisions that are closely related to strategies used by political parties and local influential individuals to secure power.

This is demonstrated by disruption of trade and the exclusion of Somali and Orma pastoralists from Endau in recent years. During the 1970s, raids by Somali pastoralist groups (that may, according to group interviews, have been politically instigated by powerful people and at the very least facilitated by those in charge of security authorities with ethnic preferences) led to deaths, migration and destitution in terms of livestock loss and landlessness among some of the Endau Kambas. More recently, killing of animals and attacks on people have been associated with internal struggles within the Endau Kamba community to control trade and economic and political influence.

In general, prolonged droughts have forced the ethnic groups to coexist peacefully due to the mutual dependence on trade and access to water and grazing. However, Eriksen and Lind (2009) describe how, in 2006, strong internal wrangling took place among Kamba politicians, elders, traders and local administrators over trade with pastoralists of Somali and Orma ethnic origin (trade with other Kamba pastoralists not being questioned). Some politicians and local leaders caused tension and created an atmosphere of conflict among the agropastoralist Kamba and pastoralist Somali people. Some leaders organized public meetings where it was decided that trade with people of Somali and Orma ethnic origin was to be discontinued. When some local Kambas continued to trade, those in favour of excluding Somali and Orma rallied youths to attack Somali camels, and local government administrators who favoured trade were suspended. Some politicians in particular influenced the office of provincial administration not to allow the Somalis to interact with Kambas peacefully. Representatives of the district and provincial administrations were brought in to arrest Kamba traders in order to stop what was presented as a conflict with pastoralists. Later, the local peace committee convinced the provincial administration to allow trade as long as Somalis and Ormas resided several kilometres away from Endau, visiting markets and wells with livestock only when necessary. The local government administration was also reinstated.

Even if no longer expressed in formal laws and policies, ethnically based restrictions on movement within Kenya have continued to exist in practical decision making, with severe consequences for pastoral–agropastoral interactions, for the capacity to adapt to climate variability and essentially for human security in terms of political and social rights. In 2007, in the run-up to national elections, ethnic Somalis and Ormas were excluded from the area entirely. Interaction between them and the local population was disrupted before the end of the dry season by the provincial administration, following a decision made at a public meeting in which people were asked to vote on whether to exclude Ormas and Somalis from the Endau area. The pastoralists had initially been rented pasture land by a group of Kamba elders, but a dispute occurred when some of the pastoralists allowed the animals to graze on farmland. In the dispute, a Kamba youth killed a camel, and one of the youths was cut with a knife. A meeting was held between the local elders, the local chief, the District Commissioner and the local politicians, including the MP.

In effect, the decision to exclude Somali and Orma pastoralists had no legal, constitutional or policy basis and was not based on consensus decision making. It is likely that the views of influential individuals dominate decision making in public meetings. Some of the people who do not own wells and those who do not transact business with the pastoral groups were in favour of exclusion. Although presented as a public decision, group and key informant interviews (traders, well owners and farmers) revealed that the majority of the members of the settled agropastoral Kambas were against the decision to stop trade. They emphasized that it was only through trade that peace and drought incomes could be ensured, but lamented that they could not declare this openly in meetings because of fear of the provincial administration. Informants described the process as a decision made by the District Commissioner and provincial administration, with the public rally serving as the forum to promulgate an order to people to stop trade. The decision was enforced by the police through public flogging and filling in of wells.

When the research team visited the study area in 2007, no Somali and Orma were seen, and key informants among farmers and administration explained that the groups were chased away after a brief stay. It is likely that this total exclusion was related to national politics, because it is common in Kenya to segregate ethnic groups geographically during elections to solidify politicians' control over voter opinion and support (most dramatically demonstrated by the post-election violence in other areas of Kenya in early 2008).

The effect was dramatic on both Kamba agropastoralists and Orma and Somali pastoralists because the pre-election period coincided with a drought. Many agropastoralist groups lost their incomes because they were dependent on trade with pastoralists. Market surveys revealed that the turnover in trade and marketing of cereals and livestock went down drastically. The local Kambas had invested in digging wells (wells in river beds have to be dug every year after the rains stop and the rivers dry up) but did not get returns on their investments because, without Somali and Orma pastoralists, the wells were not rented out for long enough periods. The pastoralist groups were also affected because they could not sell their animals in exchange for food, and they could not get water and pasture for their animals. Interviews revealed that some pastoralists moved to safe areas where there is a population of sedentary Ormas and Somalis, such as in Mutha to the south of Endau (where interaction between Kambas, Ormas and Somalis is peaceful, with some Ormas and Somalis recently settling to own shops, trade and send their children to school) and in the Tana River District (where there are traditional wet-season grazing areas).

In key informant interviews, government administrators and politicians argued very strongly that people of Orma and Somali ethnic origin belonged to other districts and should not be allowed into Kitui District without a letter from their own district administration. In 2007, the Ormas and Somalis were required to get letters from the provincial administration in their 'home' district (Tana River) to come into the area detailing their period of residence and number of livestock and people. The same requirement was not imposed on the Kambas from Mwingi who were considered part of the larger Kamba community as they have relatives or kin in the area and enjoy close social networks. In independent Kenya, the need to have permits to move from district to district has been interpreted by the pastoral groups largely as an attempt by the provincial administration to block their movement; it is not officially practised in other parts of the country.

The struggles for political influence and decision making in the drylands affect not only interactions, movement of livestock and trade as described above; related conflicts both within and between groups creating vulnerable groups as well as leading to loss of social and political rights. The effects of conflict on vulnerability are illustrated in Table 3 and further explained by Lind and Eriksen (2006). Interviews carried out in 2007 revealed that the loss of agropastoralist–pastoralist interaction in Endau has greatly disrupted the local livelihood patterns that depend on seasonal movement of livestock. At the Twambui trading centre, some shops had to close and the vibrant economic activity of earlier years has been drastically reduced, suffocating the local economy.

4. Conclusions

The experience of households in the drylands of Kenya can provide useful lessons for sustainable adaptation. The case of Endau exemplifies that flexible interactions among groups through trade, mobility and natural resource access are essential components of sustainable adaptation. However, current policies and informal governance are impeding rather than facilitating such adaptation. In particular, interventions and decisions by the government and district administration (fuelled by local leaders and politicians) that obstruct interaction and trade have serious implications for equity because sources of trade are lost and because decisions favour some individuals and ethnic groups over others. The current governance has to some extent also initiated and contributed to tension and conflicts between and within agropastoral and pastoral groups, hindering access to natural resources critical for managing drought, creating vulnerable groups as well as leading to loss of social, political and human rights. Hence, such governance has undermined sustainable adaptation and human security.

The case of Endau also shows that access to land, forest and water resources is important for the strategies of the poorest groups in society and hence for social equity in adaptation. This does not mean that strict conservation policies necessarily contribute to sustainable adaptation, however, because conservation of the hilltop in the name of environmental integrity has denied poor people access to resources, reduced their adaptive capacity and led to destitution.

Instead, policies must promote both environmental integrity and resource access, but most critically the dynamic interactions between groups and mobility in accessing resources to be able to withstand drought.

The discussion above shows that both the official policies and the state practice in terms of actual decision making (whether in line with policy and legal frameworks or not) appear to undermine human security and sustainable adaptation in Endau. Formal policies interact with power structures and political processes to create practical decisions and a governance regime that threatens sustainable adaptation in terms of social equity in adaptation options, critical pastoral–agropastoral interactions, and forest resource rights. Improvements have taken place in social and political rights in terms of democratization, increased development funding and local involvement in forest management (the latter more in policy than in practice). However, the decisions that are made in practice, regarding animal movement, pastoralist–agropastoralist interaction and forest use, undermine people's political rights and threaten adaptive capacity, hence leading to destitution and social inequity. Crucially, conservation policies treat different natural resources (such as water and forest) and sectors (such as livestock, agriculture and forestry) as isolated from each other. In addition, decisions seek to tie ethnic groups and their resource uses to specific geographic areas, rather than enhancing flexibility and interactions that are actually the basis of adaptation to climate variability and change in the drylands.

Pouliote et al. (2009) suggest that so long as the root causes of poverty remain, involving economic structures, policies and institutions, inequalities, and access to resources, the adaptations will be limited to short-term stop-gap efforts and hence cannot be sustainable. The case of Endau shows that adaptation measures must similarly address the root causes of vulnerability in order to be sustainable. In line with suggestions by Eriksen et al. (2011), this means that the vulnerability context and the multiple stressors faced by dryland populations need to be understood, such as restrictions on mobility and interaction, conflicts, exclusion from key resources for coping with drought and destitution. The second principle of sustainable adaptation identified by Eriksen et al., that is the need to recognize and mediate differing adaptation interests and outcomes, also emerges clearly from the politicization of pastoralist–agropastoralist interaction in Endau. For adaptation policies to be effective and to address social equity in particular, they need to ensure that the various production systems and adaptation needs of different groups are supported, that is, for the case of Endau the diverse activities of pastoral and agropastoral groups and the interaction and interdependence between them. The case also underscores the importance of the third principle for sustainable adaptation: to integrate local knowledge into adaptation responses. In the case of Endau, this entails that planners take into consideration the seasonal movements of people and livestock, strengthen resource rights to forest resources used to manage drought and strengthen customary natural resource management systems and institutions.

Local adaptation measures alone cannot sufficiently create sustainability. Instead, they need to be placed in the context of national and international policies and structures. In several meetings held at village, district and national levels, mobility of people and livestock (access to grazing), diversification of crops and income sources, and flexible access to forest were all identified as important in helping the agropastoral and pastoral groups adapt to climate changes (Owuor et al., 2005). Investments such as those to improve water supplies were also considered critical measures. Furthermore, enhancing equity, and hence social sustainability in adaptation, is linked to strengthening numerous social and political rights. For example, targeting landlessness and destitution, and enhancing democratization through representation of the most vulnerable populations in development and resource management forums, as well as through strengthened conflict resolution mechanisms would be important adaptation measures (Eriksen and Lind, 2009).

The case of Endau illustrates that a more sustainable adaptation pathway for the drylands would require a fundamental change in governance regime from one of imposing punitive measures that dynamic interactions to one through which, instead, interactions among pastoral and agropastoral groups are facilitated and strengthened. It also involves moving from a regime of weakening local resource rights to deliberate efforts by the government and other actors to engage the local population in decision making and benefit sharing.

Acknowledgements

The authors are grateful to all the respondents in Kitui and the communities in Endau, including the leaders, general population and local administration. The government officials at the district and divisional levels were also very helpful and provided a conducive environment in which we could work. We wish to thank our employer, the Kenya Forestry Research Institute, for allowing us to be part of this collaborative research and for all the logistical support provided for the fieldwork. Special thanks go to the Research Council of Norway for financial support through the Centre for International Climate and Environmental Research – Oslo and the University of Oslo.

Note

1. Since the time of data collection, the former Kitui District has been split up into several new administrative units.

References

Benjaminsen, T. A., 2008. Does supply-induced scarcity drive violent conflicts in the African Sahel? The case of the Tuareg rebellion in northern Mali. *Journal of Peace Research*, 45(6). 831–848.

Burton, I., Huq, S., Lim, B., Pilifosova, O. and Schipper, E. L., 2002. From impacts assessment to adaptation priorities: the shaping of adaptation policy. *Climate Policy*, 2(2–3). 145–159. doi: 10.1016/S1469-3062(02)00038-4.

Eriksen, S., Aldunce, P., Bahinipati, C., Martins, R., Molefe, J., Nhemachena, C. R., O'Brien, K., Olorunfemi, F., Park, J., Sygna, L. and Ulsrud, K. (2011). When not every response to climate change is a good one: identifying principles for sustainable adaptation. *Climate and Development*, 3(1).

Eriksen, S., Gachathi, F. N., Muok, B., Ochieng, B. and Owuor, B., 2006. Synergies in biodiversity conservation and climate change adaptation: the case of hilltop forests in Kitui, Kenya. *Putting People Back into Nature. Lessons from Savannas and Dry Forests*, J. Mistry and A. Berardi (eds). Ashgate, Aldershot, UK. 187–226.

Eriksen, S. and Lind, J., 2009. Adaptation as a political process: adjusting to drought and conflict in Kenya's drylands. *Environmental Management*, 43(5). 817–835. doi: 10.1007/s00267-008-9189-0.

Gachathi, F. N. M., 1996. The conservation priorities in the arid and semi-arid lands. *The Biodiversity of African Plants*, L. J. G. van der Maesen (ed.). Kluwer Academic Publishers, Dordrecht, the Netherlands. 313–316.

Government of Kenya, 1999. *Population Census*, Kenya Bureau of Statistics.

Government of Kenya, 2005. *Forest Act, 2005*, The Government Printers, Nairobi.

Heald, S., 1999. Agricultural intensification and the decline of pastoralism: a case study from Kenya. *Africa*, 69(2). 213–238.

IPCC (Intergovernmental Panel on Climate Change), 2007. Climate change 2007: impacts, adaptation and vulnerability. *Contribution of Working Group II to the Second Assessment Report of the Intergovernmental Panel on Climate Change*, M. Parry, J. Palutikof, P. Vanderlinden and C. Hanson (eds). Cambridge University Press, Cambridge, UK.

Kemp, R., Parto, S. and Gibson, R. B., 2005. Governance for sustainable development: moving from theory to practice. *International Journal of Sustainable Development*, 8(1/2). 12–30.

Korir, M. K., 1991. History of environmental management in Kenya. *Gaining Ground: Institutional Innovations in Land-Use Management in Kenya*, A. Kiririo and C. Juma (eds). Rev. edition, Acts Press, African Centre for Technology Studies, Nairobi.

Lind, J. and Eriksen, S., 2006. The impacts of conflict on household coping strategies: evidence from Turkana and Kitui Districts in Kenya. *Die Erde*, 137. 249–270.

McSherry, B. and Brass, J. N., 2007. *The Political Economy of Pro-Poor Livestock Policy Reform*, IGAD LPI Working Paper No. 04-08, IGAD, Nairobi.

Mortimore, M. J. and Adams, W. M., 2001. Farmer adaptation, change and crisis' in the Sahel. *Global Environmental Change*, 11. 49–57. PII: S 0 9 5 9 - 3 7 8 0 (0 0) 0 0 0 4 4 - 3.

Morton, J., 2005. *Legislators and Livestock: A Comparative Analysis of Pastoralist Parliamentary Groups in Ethiopia, Kenya and Uganda*, Final Report of the NRI/PENHA Research Project on Pastoralists Parliamentary Groups, Funded by DFID Livestock Production Programme and the Cape Unit, African Union's Interafrican Bureau of Animal Resources.

Oba, G., 2001. The importance of pastoralists' indigenous coping strategies for planning drought management in the arid zone of Kenya. *Nomadic Peoples*, 5(1). 89–118.

O'Brien, K. L., 2006. Are we missing the point? Global environmental change as an issue of human security. *Global Environmental Change*, 16(1). 1–3.

Owuor, B., Eriksen, S. and Mauta, W., 2005. Adaptation to climate change: evidence from a dryland mountain in Kenya. *Mountain Research and Development*, 25. 310–315.

Pouliote, J., Smit, B. and Westerhoff, L., 2009. Adaptation and development: livelihoods and climate change in Subarnabad, Bangladesh. *Climate and Development*, 1(1). 31–46.

Reid, P. and Vogel, C., 2006. Living and responding to multiple stressors in South Africa – Glimpses from KwaZulu-Natal. *Global Environmental Change*, 16(2). 195–206. doi: 10.1016/j.gloenvcha.2006.01.003.

Republic of Kenya, 2003. Geographic dimensions of well-being in Kenya: Where are the poor? Vol. 1. Central Bureau of Statistics. Ministry of Planning and National Development, Nairobi.

Schubert, R., Schellnhuber, H. J., Buchmann, N., Epiney, A., Griesshammer, R., Kulessa, M., Messner, D., Rahmstorf, S. and Schmid, J., 2008. *Climate Change as a Security Risk*. Earthscan, London.

Scoones, I., 1995. *Living with Uncertainty: New Directions in Pastoral Development in Africa*. Intermediate Technology Publications, London, UK.

Selboe, E., 2008. *Changing Continuities: Multi-Activity in the Network Politics of Colobane, Dakar*. PhD thesis, Department of Sociology and Human Geography, University of Oslo, Norway.

Smit, B., Burton, I., Klein, R. J. T. and Wandel, J., 2000. An anatomy of adaptation to climate change and variability. *Climatic Change*, 45(1). 223–251.

Thébaud, B. and Batterbury, S., 2001. Sahel pastoralists: opportunism, struggle, conflict and negotiation. A case study from eastern Niger. *Global Environmental Change*, 11(1). 69–78.

Thomas, D. S. G., Twyman, C., Osbahr, H. and Hewitson, B., 2007. Adaptation to climate change and variability: farmer responses to intra-seasonal precipitation trends in South Africa. *Climatic Change*, 83(3). 301–322.

Tschakert, P., 2007. Views from the vulnerable: understanding climatic and other stressors in the Sahel. *Global Environmental Change*, 17. 381–396. doi: 10.1016/j.gloenvcha.2006.11.008.

Twyman, C., Sporton, D. and Thomas, D. S. G., 2004. Where is the life in farming?: the viability of smallholder farming on the margins of the Kalahari, Southern Africa. *Geoforum*, 35. 69–85. doi: 10.1016/S0016-7185(03)00030-7.

Ziervogel, G., Bharwani, S. and Downing, T. E., 2006. Adapting to climate variability: pumpkins, people and policy. *Natural Resources Forum*, 30(4). 294–305. doi: 10.1111/j.1477-8947.2006.00121.x.

climate and development

Gums and resins: The potential for supporting sustainable adaptation in Kenya's drylands

FRANCIS N. GACHATHI[1,*] and SIRI ERIKSEN[2]

[1]Kenya Forestry Research Institute, P.O. Box 20412-00200, Nairobi, Kenya
[2]Department of International Environment and Development Studies – Noragric, Norwegian University of Life Sciences, P.O. Box 5003, No-1432 Aas, Norway

Frequent droughts and conflicts are key challenges faced by nomadic pastoralists in Kenya's drylands. Few options exist for alternative livelihoods. This article investigates the potential of collecting plant gums and resins for livelihood diversification and for contributing to sustainable adaptation to climate change in Kenya's drylands. Observations were made in various studies over a period of two years on dryland vegetation resources, and interviews undertaken with gum collectors. This research found that many households currently collect and sell plant gums and resins as alternative to livestock production. These include gum arabic from *Acacia senegal* and *Acacia seyal*; myrrh from *Commiphora myrrha*; hagar from *Commiphora holtziana*; and frankincense from *Boswellia neglecta*. Collectors include poor people, women and children and some opportunists. Incomes are relatively low, however, and several factors constrain the activity. The case of gums and resins illustrates that key principles of sustainable adaptation are related: supporting local knowledge and adaptation strategies (one key principle of sustainable adaptation) does not contribute to sustainability unless at the same time contextual factors that marginalize livelihoods are addressed (another key principle).

Keywords: adaptation sustainability; drought; drylands; gums and resins; livelihood diversification

1. Introduction

Human survival in the Kenyan drylands is continuously faced with multiple challenges, including frequent droughts and unpredictable cattle rustling and violence. Generally, droughts affect the quality and quantity of water and grazing resources, reduce incomes and aggravate poverty. They may trigger conflicts, cattle raids, loss of life in both humans and livestock and accelerate the rate of land degradation. Conflicts, including banditry and raids, increase pastoralists' vulnerability and hinder human and animal movement (Eriksen and Lind, 2009). Despite these predicaments, drylands are very rich in biodiversity and are important areas for livestock production. The harsh physical environment in the drylands directly support the livelihoods of close to 25% of Kenya's population and slightly more than 70% of the livestock population, whereas 90% of the gazetted national parks and game reserves are located in the drylands (KEFRI, 1992). Although many dryland communities in Kenya continue to rely largely on livestock for their livelihoods, there has been a process of marginalization of pastoralism and a related sedentarization that has increased the vulnerabiliy of these populations (Mkutu, 2007). These processes have come about through a combination of policies and development interventions that have been aimed at settling populations and incorporating them into non-livestock economic activities such as farming or fishing, and livestock losses due to conflict and

■ *Corresponding author: *E-mail:* gachathif@yahoo.com

ISSN: 1756-5529 (print)

drought (Hogg, 1987; Heald, 1999; Owuor et al., 2011). Insecurity and a lack of veterinary and other livestock support services have meant that it is increasingly difficult to recover and restock after livestock losses, leaving an increasing number of people with small herds or no herds at all (Eriksen and Lind, 2009).

Diversification of livelihoods is frequently cited, both in research and policy, as a main way to cope with and reduce vulnerability to drought (Scoones et al., 1996; Republic of Kenya, 2003). Climate change makes this issue even more pertinent. The local manifestations of climate change are not easily predictable because in East Africa warming is likely to lead to drying in some areas and increased precipitation in others (Boko et al., 2007). However, seasonal shifts in rainfall and higher temperatures associated with climate change mean that growing uncertainty and incidences of prolonged droughts are challenges with which populations may increasingly have to cope (Hulme et al., 2001). The need to alter drought grazing patterns at the same time as insecurity and development policies render many areas inaccessible reinforces the vulnerability of pastoral livelihoods.

The collection of gums and resins illustrates how sustainable adaptation to climate change could potentially be supported – it represents an alternative source of income for poor people during drought and as a non-wood forest resource it is likely to be environmentally sustainable. Strengthening the local adaptation options of pastoral communities requires recognition of the opportunities that the natural resources of dryland ecosystems, provide for diversifying income sources. Plant gums, gum arabic and plant gum resins, myrrh, hagar and frankincense are among these natural resources. One advantage of these resources is their ability to produce in the dry season. Increased exploitation of indigenous gums and resins represents an opportunity to build adaptation on local knowledge, a key principle of sustainable adaptation identified by Eriksen et al. (2011); however, the exploitation of these resources comes with new challenges. This article critically explores whether exploitation of gums and resins can contribute to sustainability in terms of both social equity and environmental integrity. Data on the collection and sale of gums and resins collected between 2006 and 2008 are analysed in order to establish the scale and methods of collection, which groups participate in the collection and the extent to which these resources contribute to incomes. This analysis suggests that harvesting gums and resins could provide a viable option to strengthen livelihoods in a manner that is environmentally sustainable. However, the activity is currently constrained by a range of factors closely associated with the general marginalization of drylands, that is, lack of marketing and infrastructure, poverty, policy neglect, conflict and insecurity. The case of gums and resins illustrates that key principles of sustainable adaptation are related: supporting local knowledge and adaptation strategies (one key principle of sustainable adaptation) does not contribute to sustainability unless at the same time contextual factors that marginalize livelihoods are addressed (another key principle).

2. Background: Kenya's drylands

Kenya's drylands, defined as areas receiving annual rainfall below 1,000 mm, are characterized by a hot, dry climate and scant vegetation cover. The rains are not only low but erratic, and typically of short duration but often of high intensity and therefore highly erosive. Recurrent drought and occasional floods are also features of the drylands. The soils are generally poor, with high sand content, poor surface structure due to erosion and often saline with low organic matter. The drylands cover about 80% of total land area (Figure 1). Most of the country's wood-fuel-based energy requirements are met from drylands. Conflict over resources and household food insecurity are common, while the incidence of poverty is the highest in Kenya (Government of Kenya, 2000). The most suitable land use is livestock keeping and wildlife conservation.

Traditionally, nomadic pastoralism is the mainstay of the drylands economy. Vegetation

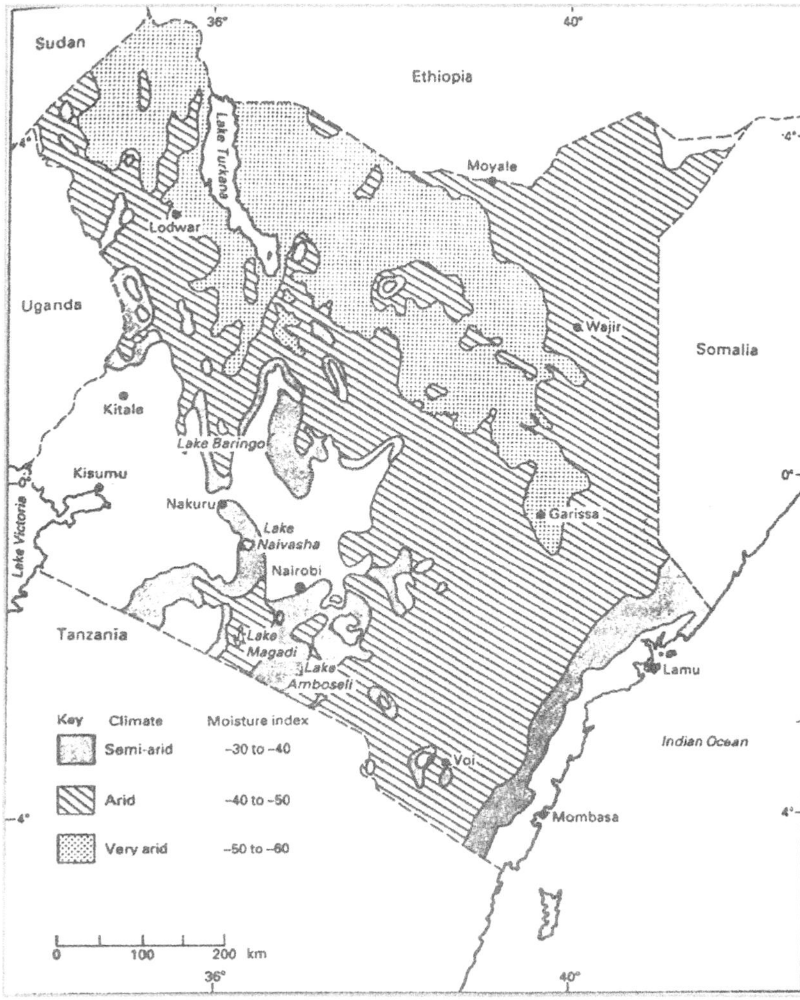

FIGURE 1 A map of Kenya's drylands showing semi-arid, arid and very arid areas and neighbouring countries

is thus vital to the pastoral communities for grazing and browsing. The pastoral production system operates by independent family units exploiting a common resource (e.g. water, grazing land) through privately owned and managed herds. Frequent daily movement of herds between pastures helps prevent overuse of a single area's biomass and helps avoid disease in their livestock. Herd diversification (cattle, camels, sheep, goats and donkeys) ensures that both browsers and grazers are present and that an extended dry spell or a single disease is less likely to wipe out an entire herd. This is a highly flexible system, which has evolved over time as the most efficient means of exploiting transient water and pasture under ecologically marginal conditions, available technologies and the prevailing economy (Chikamai and Odera, 2002).

Frequent, prolonged and severe droughts are posing great challenges to survival in Kenya's drylands. Dams and wells are drying up and loss of human life and livestock is increasingly frequent. Droughts cause food shortages, affect the quality and quantity of water resources, reduce incomes and aggravate poverty in the region. The vulnerability of pastoral livelihoods is exacerbated by conflict and insecurity. The causes of conflict are complex, related both to strategies to acquire wealth (cattle), political developments, marginalization, the failure of the state to provide security infrastructure and the loss of – and search for – dry season pastures (Eriksen and Lind, 2009). As pastoralists migrate to areas where they can access pasture and water, bloody confrontations have occurred between various pastoralist communities and among clans of the same community. For instance, on 15 September 2009 *Daily Nation*, a popular national newspaper, reported that 31 people were killed in a cattle rustling battle between community members of the Samburu and Pokot. Conflicts have even crossed country borders creating serious diplomatic feuds between Kenya and its neighbours. On 6 October 2006, *The Standard* newspaper, another popular national daily, reported Kenya police had shot dead 16 Ethiopian cattle rustlers and a Kenyan woman. The northern region of Kenya borders parts of Somalia, Southern Sudan, Southern Ethiopia and Eastern Uganda (Figure 1), where the proliferation of small arms among pastoralists is a major security concern.

In this context of vulnerability and loss of pastoral livelihoods, the harvesting of gums and resins could provide a much needed source of income. At the same time, the same contextual factors with which livestock keeping is faced, including conflict and poor infrastructure, may also constitute challenges for gums and resin harvesting and its ability to support flexible pastoral livelihoods. Investigating the role of gums and resins in current livelihoods is important in order to enhance our understanding of the potential and limits to enhancing social equity and environmental integrity – that is, sustainable adaptation – through livelihood diversification and increased exploitation of indigenous vegetation resources.

3. Methods and study areas

This article presents research carried out between 2006 and 2008 on dryland vegetation resources. The study applied participatory approaches involving a semi-structured questionnaire, and interviews covering a checklist of questions related to various vegetation resources, including gums and resins. This process analysed the socio-economic status and conditions of the rural community through sharing, planning and exchanging of ideas in order to identify problems. The method focused on local communities and local traders in districts in northern Kenya where gums- and resins-producing tree species were abundant and the collection and marketing of gums and resins was well defined. These districts include Isiolo, Garissa, Wajir, Mandera, Moyale, Marsabit and Samburu. Areas with high levels of distribution, collection and marketing of these products (Beentje, 1994; Chikamai and Gachathi, 1994; Gachathi, 1994; Curry, 2000; Chikamai, 2002; Chikamai and Odera, 2002; authors' research) and local communities were given preference.

In the field, a random sampling design was used in the sample plots to determine densities of the resources. The sample plots were in part determined by accessibility, and the approximate location and the total number of the plots were determined by the size of the area of occurrence of the resource; that is, the larger the area covered, the more sample plots. Sample plots of 20 m × 20 m (0.04 ha) were established using a 30 m linear tape. Details of the plots' GPS location were recorded in the field data sheets. Both the botanical and local names of the resources being assessed were recorded. Other details taken in each sample plot were:

- number of mature trees (trees that were producing gums or resins)
- number of juvenile trees or saplings (those that were too young to produce gum)
- soil type (sandy, rocky, loam, black cotton, etc.)
- extent of the resource and names of the collection centres in the area, and
- names of members of the local community interviewed at the site and their contacts.

GIS mapping techniques were applied in generating distribution maps of the resources.

Socio-economic information was collected with the mutual understanding that personal information would be confidential. Personal details included sex, age, marital status and household income. The general objective was to discover the different types of resources traded, the types of collector, the traders and the entire chain through to exporters, to understand the constraints and opportunities in the gums and resins sub-sector. In addition, the socio-economic information was intended to capture the effects of environmental and climate variability and change, including through water shortage, death of animals and increased conflict. Other information included gum collection records, post-collection handling, utilization and marketing. Respondents also provided data on seasonality and distribution of the resources. About 200 members of the local population were interviewed alongside gum collectors, traders and their agents.

4. Findings

4.1. The use of gums and resins resources

Kenya has significant resources of gum arabic and gum resins, with commercial production confined to the north of the country (Figure 2). Trees and shrubs that dominate the drylands and form an important component of the vegetation are often referred to as *Acacia-Commiphora* woodland (Beentje, 1994). Traded gum arabic is from *Acacia senegal* (L.) Willd. var. *kerensis* or *Acacia seyal* Del. var. *seyal*. Gum resins are myrrh from *Commiphora myrrha* (Nees) Engl., oppoponax or hagar from *Commiphora holtziana* Engl. and frankincense from *Boswellia neglecta* S. Moore.

The study revealed that the production of gums and resins is relatively widespread. A total of 198 temporary sample plots were marked in all areas where the resource occurred, which were used to assess densities and distribution. *A. senegal* was distributed throughout the seven districts, with densities ranging from 975 to 1,008 trees/ha. *A. seyal* was found in Moyale, Wajir and Mandera where densities of 278 and 277 trees/ha were recorded. *C. myrrha* was confined to Wajir and Mandera and had densities ranging between 100 and 83 trees/ha. *C. holtziana* occurred in all districts with densities ranging between 221 and 154 trees/ha. *B. neglecta* was found in all the districts with densities ranging between 600 and 95 trees/ha. Annual production estimates were: 2,000 tonnes for gum arabic; 2,500 tonnes for hagar; 900 tonnes for myrrh; and 100 tonnes for frankincense.

Interview data suggest that the collection of gums and resins is an activity largely carried out by pastoralist peasants who can be classified as follows:

- Poorest of the poor in the community. These are people who either have no livestock or have very few animals left. Mostly they or their parents have lost livestock due to raids through banditry, rustling, drought or disease or when a father dies and the children are left as orphans or without proper care.
- Women and female children as they fetch water, firewood or tend livestock.
- Emerging gums and gum resin opportunists, that is, those cashing in on emerging opportunities in the sub-sector in addition to their normal livelihoods activities.

Gum and resin harvesting appears to represent a source of income for many of the poorest and vulnerable groups (in addition to emerging opportunists) and may therefore contribute to alleviating poverty. The extent to which this activity is able to counteract social inequity (and support sustainable adaptation) depends on the levels of income, in particular the potential to provide an income high enough to meaningfully compensate or substitute that of livestock herding, or to assist people back into higher-income livelihoods, for example, through enabling restocking. The levels of collection,

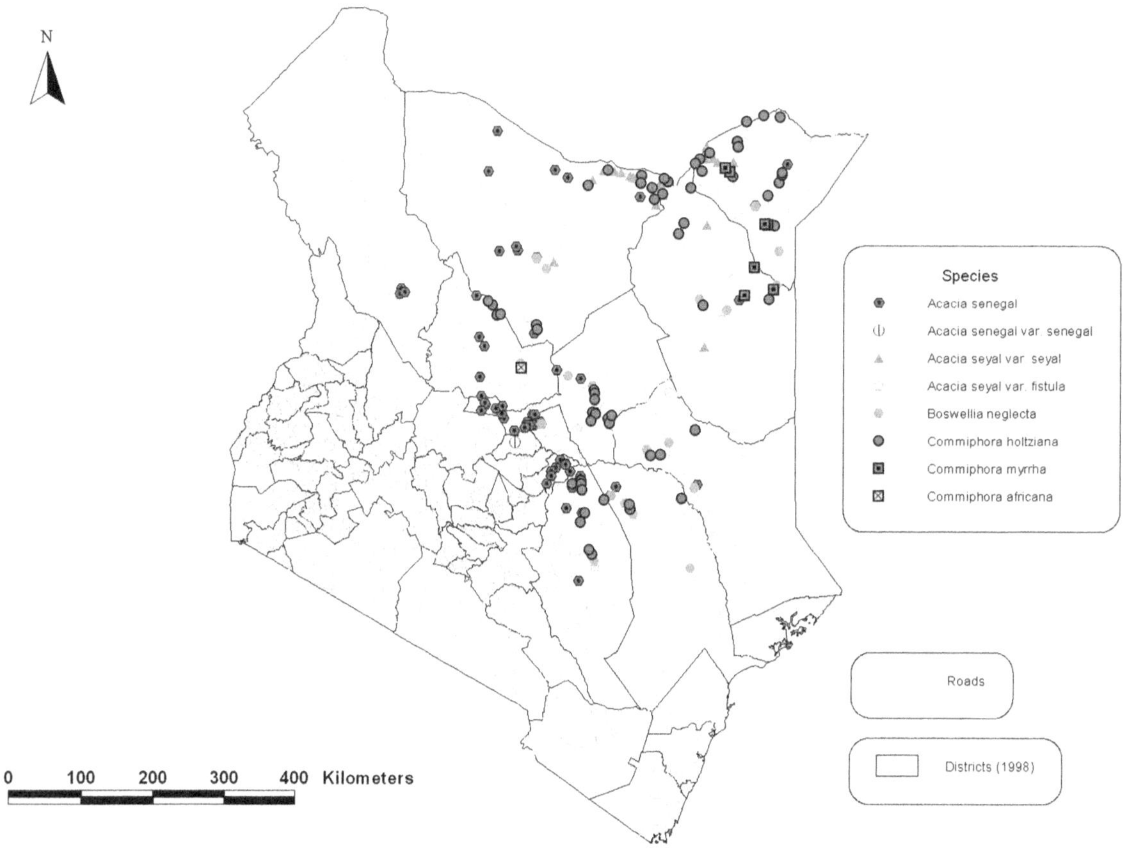

FIGURE 2 Distribution of gum arabic and gum resin resources

trade and incomes from this activity are investigated below.

4.2. Gum arabic: description, collection and trade

In its natural state gum arabic comes in a variety of shapes, colours and sizes. The colour of the gum may vary from colourless through different shades of yellow, amber, orange, red and dark brown. The best grades from *A. senegal* are in the form of whole, round 'tears', orange-brown in colour and with a matt surface texture. Gum from *A. seyal* is more friable than that produced by *A. senegal* and is rarely found as whole lumps. Use of gum arabic falls into three main sectors: the food industry, the pharmaceutical industry and industries such as printing, ceramics and textiles (Chikamai and Odera, 2002). Locally, it is eaten as food and has some medicinal uses.

The local population exploits the resource for both subsistence and commercial purposes. Subsistence exploitation, which involved collection of gum arabic for use as food, is common in times of food scarcity and when herding livestock. Commercial exploitation involved collection of gum arabic for sale in cash or in exchange for commodities including foodstuffs.

Gum arabic is not tapped but usually collected from natural exudation or other forms of injury. The gum is removed by hand, or if hard, knocked off by a machete or stick, occasionally dropping on the ground and getting contaminated with soil. No special container is used although most collectors use plastic or gunny bags (burlap sacks). The amount collected varies depending on the collector and the season.

During a good gum season a collector can collect an average of 5 kg in a day. Allowing for other family activities, a collector may work an average of four days a week and the total volume collected in a month be as high as 100 kg. The gum is brought home and kept in similar but bigger containers while awaiting sale. The gum is sold either to local agents in remote areas in the bush or traders in local trading centres. Two modes of sale are commonly used: barter, which involves exchange of gums for foodstuffs (e.g. sugar, maize flour, rice, beans, tea leaves, salt) or other goods (beads, tobacco, human or livestock medicines) and cash. Prices vary greatly, from a peak of KES120 per kg in 1994 to an average of KES35 per kg in 2010. This translates to KES 4,000 per collector per month. Costs incurred by a collector are mostly in kind. Table 1 presents a summary of the activities in production and income.

4.3. Myrrh: description, collection and trade

Myrrh is the gum–resin exudate from the stems of *C. myrrha*. It oozes and hardens to form lumps of varying shapes and sizes of variable colour from red, brown to dark brown. Locally, it is used to make ink used in Quranic schools, for burning to repel snakes and offensive insects and in medicine to treat various ailments. Its commercial uses are mainly essential oils, cosmetics, flavourings, antiseptics and other medicines.

Myrrh is collected after the trees are tapped. This is done by organized groups of men called malmaley (malmal collectors). They live in camps each covering about 4 km^2 where each person is allocated an area with about 200–300 trees and can tap up to 100 trees per day. A specially designed axe is used for tapping and collecting myrrh. A single malmaley can collect up to 5 kg per day during the pick season. As they are in the camp throughout the tapping season, collection is a continuous process and a person can collect up to 150 kg in a month. Myrrh is collected in a tin or plastic container and stored at a collection centre in gunny bags with a polythene lining to prevent loss of volatile components. Myrrh is sold to wholesalers who also own general shops. Prices of myrrh ranged between KES80 and KES120 per kg. Taking an average of KES100 per kilo, a malmaley earns up to KES15,000 per month. Costs incurred are either in kind or foodstuffs, medicines, tapping and storage equipment, which account for up to 3,500 per malmaley per month. A malmaley therefore has a net income of about KES11,500 per month. Table 1 presents a summary of the activities in production and income.

TABLE 1 A summary of activities and income at collection stage (100 KSh is approximately 0.75 GBP)

Activity	Gum	Myrrh	Hagar	Frankincense
Average amount collected per day/person (kg)	5	5	8	4
Average monthly collection (kg)	100	150	160	80
Operational costs (labour, transport, etc.)	Kind	3,500	Kind	Kind
Average sale price (KES/kg)	35	100	45	35
Gross income per month per person (KES)	3,500	11,500	7,200	2,800

4.4. Hagar (opoponax): description, collection and trade

Hagar is oily resin exudate from the stems of *Commiphora holtziana*. It oozes out and hardens to form lumps of various sizes and shapes with variable colour from yellow to dark brown or black. Locally, hagar is used as acaricide against ticks, snakebites, scorpions, foot rot, mange and other livestock ailments. Commercially, it is a well-established herbal medicine, and used in essential oils and cosmetics.

Hagar is tapped in the same way as myrrh. But unlike myrrh, which is restricted to Wajir and Mandera districts, hagar has a very wide distribution and has the highest volume of commercial gum resins traded in northern Kenya. In addition, women are also involved in collection of hagar. Experienced tappers would collect up to 8 kg per day in the pick season and the total volume collected in a month could go up to 160 kg. Prices of hagar ranged from KES30 to KES60. At an average of KES45 per kg, a collector earns KES7,200 per month. Table 1 presents a summary of the activities in production and income.

4.5 Frankincense: description, collection and trade

Frankincense is the exudate from the stems of *B. neglecta*. It oozes out in small droplets that harden to form nodules or large lumps. It is of two types, black and white. Locally, it is used as chewing gum, burnt as incense, as perfume and as medicine for a wide range of ailments. Commercially, it is used for essential oil in the perfumery, cosmetics and the flavouring industry.

Collection of frankincense is done by women and children. Frankincense is largely marketed locally. Collection is tedious as much of it is gathered from the ground. During the study period, an average collector would collect about 4 kg per day or 80 kg in a month. Prices of frankincense ranged from KES20 to KES50 per kg, which translates to KES2,800 per month. Table 1 presents a summary of the activities in production and income.

The investigation reveals that the income from gum and resin collection can be substantial, but is probably fairly low for many. Incomes are highly gendered since the highest income activity, myrrh collection, is carried out mainly by men. The results indicate that at present, gum and resin collection cannot substitute livestock herding as a viable and profitable livelihood option. Nevertheless, the collection only takes place during the dry season and drought hence providing an important additional income and may in principle alleviate the need to sell off livestock. Clearly, many of the collectors were poor, but the data cannot easily reveal whether gum and resin collection is an indication of a failed pastoralist and lack of decent income options or an option to rebuild livelihoods. Only some of the members of the first category of collectors (poor people in the community) could potentially directly use the proceeds to restock since the other categories, including women and children and emerging opportunists were not livestock owners. Since a head of cattle costs several thousand Kenyan shillings (the exact price depends on the state of the animal and the season), a collector would have to be at the higher end of the income range and save most of a month's earnings to be able to buy a head of cattle (although small stock such as goats and sheep are cheaper and hence more easily acquired). The extent to which restocking actually took place in practice, directly or indirectly through the incomes of men and women collectors, could be a topic for future research. The next section explores some of the constraints and opportunities for making the activity an economically and environmentally viable activity, contributing to social equity and sustainable adaptation.

4.6. Constraints and opportunities

A number of constraints were identified in the gums and resins sub-sector. Generally, there was a lack of sound market information to guide opportunities, trends, prices and general quality requirements and there were no reliable buyers and/or local markets. This was more serious particularly when a buyer was 'on and off', usually with different prices for the same commodity. The price of gums and resins was also highly variable. Interviews with gum merchants revealed that trade in gum arabic in particular greatly depended on world demand and supply. There were low volumes of production in the major producing countries of Sudan and Nigeria in 1992/93 and the resulting shortage led to an increase in prices and a search for potential production areas in other countries. Consequently, the prices soared in 1994/95 stimulating high production among Kenyan producers, which in turn led to plummeting prices the following year.

There was lack of operational capital to both collectors and traders. The collectors required cash to meet their basic needs and to be able to fully engage in collection. Equally, traders required enough available capital to buy all gums and resins from collectors. Often they lacked such capital, and it was observed that most traders used some form of barter trade using groceries in exchange for gums and resins. The issue of poor-quality gum and gum resins was also significant. Reduced quality resulted from harvesting, cleaning and handling practices; adulteration by either accidental or deliberate mixing of various gums or resins by the collectors or traders; and collection of immature products. This contributed to the fall in prices after the 1994 peak. Collectors lacked experience on tapping and handling of gums. There was a lot of damage to some of the resources, especially *C. myrrha*, through poor and excessive tapping.

These products were collected in the wild and on communally owned land. There was no clear government policy in support of production and marketing of gums and resins in the country or on access and use of dryland resources. Insecurity caused by bandits and cattle rustlers was a major concern and limited the extent to which gums and resins could be fully exploited. Dangerous wild animals also posed problems, including snakes. Lack of access roads and infrastructure/facilities and inadequacy of transport to potential production areas made mobilization of collectors, collection and transportation of harvested gums and resins very difficult. These resources are found in hot and dry areas. Drinking water requirement for the collectors was very high, but drinking water is very scarce in these areas, contributing to inefficiency of the collectors. The activity is affected by threats to the trees from which gums and resins are harvested. Recurrent droughts as experienced in 1998, 2000, 2004 and 2008 greatly affected gum production and many trees died. Tree damage through clearing and de-branching by local people for fencing *bomas* (homesteads) resulted in decreased production. Natural regeneration was often affected largely by livestock, especially goats through grazing and trampling and humans through burning. Gum and resin collection is hence limited as a livelihood option by a combination of market factors, poverty, conflict, policy neglect and environmental factors. The same development shortcomings that contribute to the marginalization of dryland populations in general also constrain the potential of gum and resin collection to form part of sustainable adaptation.

Despite these limitations, the previous section showed that gums and resins provide a valuable source of income for many poor households and casual labourers as well as businessmen in Kenya's drylands, while creating jobs for the end-user industries, as thousands of people make use of these natural products in their everyday life. Drylands are characterized by high poverty levels and food insecurity. Livelihood options are limited and the majority of the population are classified as poor. Gum and resin collection offers potential to diversify the local communities' income base and hence improve their livelihoods. There are some efforts to exploit this potential. The Gum Arabic and Resin Association (GARA) was formed to coordinate and spur the development of the industry in the country. GARA is a useful body for advising the government on formulating enabling policies; producers on sound production and quality control practices; and traders on good business practice for the benefit of the domestic and export markets. There are also training programmes for collectors run by the Kenya Forestry Research Institute (KEFRI) aimed at promoting activities for the development of gums and resins in drylands. However, these efforts cannot fully compensate for the lack of policy and infrastructural support in the drylands to date.

There is great potential in the cultural and social knowledge inherent in gum and resin collection, both in terms of social and environmental sustainability. The majority of the northern Kenya dryland inhabitants are (Kenyan) Somalis and the Oromo group, the majority being Muslims. These groups have a strong attachment to gums and resins, which are used for subsistence as well as commerce. They value these resources and have conserved them over long periods. The Somali community in particular traded these products through the former Somalia Republic. With the collapse of the Somalia Republic in the 1990s, most dealers relocated to Nairobi, helping to widen the trade of gums and resins in the country. All these players had a rich history in and knowledge of gums and resins, presenting opportunities for increased collection and local demand for the products. Efforts to strengthen local knowledge and indigenous plant-based activities could be linked to international programmes. International conventions on environment, especially the Convention on Biological Diversity and Convention on Combating Desertification, indirectly support development activities in the gums and resins sub-sector. Programmes developed in the light of these conventions stand a better chance of accessing funds from various international donors.

5. Discussion and conclusions

There is potential for gum and resin collection to contribute to social equity and environmental integrity and hence support sustainable adaptation. This potential is largely underused, however. Plant gums and resins fall into the category of non-wood forest products. They are renewable resources that could be sustainably exploited for household income and still conserve biological diversity and ecosystem functions while increasing the overall productivity of the land. This requires addressing the market, poverty, insecurity and policy factors currently constraining incomes as well as improving some of the tapping practices currently leading to damage to trees. Sustainable forest use would guarantee that goods and services could continue to be provided from the forest ecosystem under environmental, social and economic parameters. They could serve as raw materials for enterprises based on them, thus generating employment and raising the socio-economic status of poor populations (FAO, 1995). Forest investments based on non-wood products could even be more efficient and resilient, particularly if certification is included. This is especially important to Kenya's drylands since there exist few alternatives for supporting livelihood because of the difficult environmental conditions. The conditions have been worsened by increased recurrent drought and occasional flooding that has killed many livestock; the main economic, cultural and social support for the local population. One advantage of gum and resin resources is their ability to produce marketable products in the dry season when forage is scarce. The potential prospects for these resources would be even greater if they were to be properly developed (NAS, 1979; Stiles, 1988).

Gum and resin collection as a livelihood option is consistent with many policy goals. However, policy implementation has failed to support the activity in practice. Woodlands producing gums and resins have the potential to address climate change through carbon sequestration, or fixing of carbon. Among the many local environmental services they provide are soil conservation (i.e. protection against erosion and maintenance of fertility), shelter against wind and shade. Vision 2030 is a policy document that aims to launch Kenya's economic recovery strategy to bring the country up to a middle-income status with an average annual growth rate of 10%. The strategy recognizes the potential of the drylands to supply marketable commodities on a sustainable basis, among them, gums and resins. The Poverty Reduction Strategy Paper (PRSP) covering the period 2001–2004 puts a lot of emphasis on embracing environmentally sustainable development. It highlights the causal links between poverty and environment, health, government policies and institutional failures. However, the government has so far failed to implement the PRSP as per the action plan (SIDA, 2003). Environmental benefits of forest in the drylands have long been underemphasized or ignored by planners and decision-makers because they are very hard to value, and many lie far outside formal markets and pricing mechanisms (Republic of Kenya, 2003).

Successful development calls for recognizing the potential of dryland ecosystems and key commercial natural resources, supporting institutional structures and opportunities for diversifying income sources, a transition to a cash economy, creation of employment and sustainable management of the resource base. The role of gums and resins is one component of a number of non-wood forest products, which if integrated and balanced with other opportunities for dryland management, can contribute to the economic well-being and long-term viability dryland areas. There is, however, a need to understand local knowledge, land-use systems and social structures in order to be able to assist in adapting and strengthening them. As all gum and resin collection is from communally owned land, land tenure change, for example, has crucial impacts on land use and conflicts.

This study illustrates that local knowledge-based activities do not automatically lead to viable incomes and sustainable management of the resource base. However, if some of the constraints identified above can be addressed, plant gums and gum resins remain a potentially important option for livelihood diversification in Kenya's drylands, which could complement livestock production and alleviate vulnerability associated with conflict, livestock loss, drought and marginalization of pastoralism. The sub-sector has a positive impact on the economy at individual, household, community and even at national level. However, strengthening the development of this sub-sector requires a wide range of measures. Areas to emphasize include: human resource development, community participation, marketing of resources, infrastructure development, promotion of micro-enterprises and research.

Many of the constraints to gum collection identified in this study are related to poverty and a lack of infrastructure. This observation suggests that some of the contextual factors leading to the marginalization of livestock keeping are also limiting gum and resin harvesting as a livelihood option. Clearly, contextual factors creating social vulnerability among dryland populations need to be addressed if gum and resin collection is to successfully contribute to sustainable adaptation. This has important implications for how we understand the principles of sustainable adaptation identified by Eriksen et al. (2011); that is, adhering to one of the four principles alone is unlikely to successfully lead to socially equitable and environmentally sound forms of adaptation. Instead, the principles are closely related and interdependent and need to be addressed together. In the case of gum and resin collection the third principle – building adaptation on local knowledge – will not contribute to sustainable adaptation unless the first principle – contextual factors and multiple stressors – is also addressed.

Acknowledgements

Field work was facilitated by various institutions including: Kenya Forestry Research Institute (KEFRI), Ewaso Ng'iro North Development Authority (ENNDA) and Kenya Gums and Resins Limited (KGARL). Linus Wekesa, Abdi Somo, David Maitha and Paulo Van Breugel made invaluable contributions.

References

Beentje, H., 1994. *Kenya Trees, Shrubs and Lianas*. National Museums of Kenya, Nairobi.

Boko, M., Niang, I., Nyong, A., Vogel, C., Githeko, A., Medany, M., Osman-Elasha, B., Tabo, R. and Yanda, P., 2007. *Climate Change 2007: Impacts, Adaptation and Vulnerability. Contribution of Working Group II to the Fourth Assessment Report of the Intergovernmental Panel on Climate Change*, M. L. Parry, O. F. Canziani, J. P. Palutikof, P. J. van der Linden and C. E. Hanson (eds). Cambridge University Press, Cambridge, UK. 433–467.

Brenan, J. P. M., 1983. Manual on taxonomy of four species of Acacia. FAO, Rome, 11–19.

Chikamai, B. N. and Gachathi, F. N., 1994. Gum and resin resources in Isiolo District, Kenya: ethnobotanical and reconnaissance survey. *East African Agricultural and Forestery Journal*, 59(4). 345–352.

Chikamai, B. N. (ed.), 2002. *Review and Synthesis on the State of Knowledge of Boswellia Species and Commercialization of Frankincense in the Drylands of Eastern Africa*. Kenya Forestry Research Institute, Nairobi, Kenya.

Chikamai, B. N. and Odera, J. A. (eds), 2002. *Commercial Plant Gums and Gum Resins in Kenya: Sources of Alternative Livelihood and Economic Development in the Drylands*. Executive Printers, Nairobi, Kenya.

Curry, P., 2000. *Commercial Gums and Resins: Botanical Sources and Quality in Wajir and Mandera Districts*. SALTLICK, Isiolo and VSO, Nairobi.

Eriksen, S. and Lind, J., 2009. Adaptation as a political process: adjusting to drought and conflict in Kenya's drylands. *Environmental Management*, 43. 817–835.

Eriksen, S., Aldunce, P., Bahinipati, C. S., D'Almeida Martins, R., Molefe, J. I., Nhemachena, C., O'Brien, K., Olorunfemi, F., Park, J., Sygna, L. and Ulsrud, K., 2011. When not every response to climate change is a good one: Identifying principles for sustainable adaptation. *Climate and Development*, 3(1).

FAO, 1995. *Non-Wood Forest Products for Rural Income and Sustainable Forestry*. FAO, Rome.

Gachathi, F. N. M., 1994. Variation in *Acacia senegal* and its relationship with *Acacia circummaginata* and *Acacia thomasii* in Kenya. Msc thesis in Pure and Applied Fungal Taxonomy. University of Reading, Reading, MA.

Government of Kenya, 2000. Interim Poverty Reduction Strategy Paper, 2000–2003. 45.

Heald, S., 1999. Agricultural intensification and the decline of pastoralism: a case study from Kenya. *Africa*, 69(2). 213–237.

Hogg, R., 1987. Settlement, pastoralism and the commons: the ideology and practice of irrigation development in Northern Kenya. *Conservation in Africa: People, Policies and Practice*, D. Anderson and R. Grove (eds). Cambridge University Press, Cambridge, UK. 293–306

Hulme, M., Doherty, R. M., Ngara, T., New, M. G. and Lister, D., 2001. African climate change: 1900–2100. *Climate Research*, 17. 145–168.

Kenya Forestry Research Institute (KEFRI), 1992. *A Dryland Forestry Handbook for Kenya*. Kenya Forestry Research Institute, Nairobi.

Mkutu, K. A., 2007. Small arms and light weapons among pastoral groups in the Kenya–Uganda border area. *African Affairs*, 106. 47–70.

National Academy of Sciences. 1979. *Tropical Legumes: Resources for the Future*. NAS, Washington, DC. 278–291.

Owour, B., Mauta, W. and Eriksen, S., 2011. Strengthening sustainable adaptation: examining interactions between pastoral and agropastoral groups in dryland Kenya. *Climate and Development*, 3(1).

Republic of Kenya, 2003. *National Policy for Sustainable Development of the Arid and Semi Arid Lands of Kenya, Draft*. Government Printer, Nairobi.

Scoones, I., Chibudu, C., Chikura, S., Jeranyama, P., Machaka, D., Machanja, W., Mavedzenge, B., Mombeshora, B., Mudhara, M., Mudziwo, C., Murimbarimba, F. and Zirereza, B., 1996. *Hazard and Opportunities: Farming Livelihoods in Dryland Africa: Lessons from Zimbabwe*. Zed Books and International Institute for Environment and Development, London, UK. 288.

SIDA, 2003. *Country Strategy Paper 2004–2008*. Unpublished Report.

Stiles, D., 1988. Arid plants for economic development and desertification control. *Desertification Control Bulletin*, 17. 18–21.

research article

climate and development

The discourse of adaptation to climate change and the UK Climate Impacts Programme: De-scribing the problematization of adaptation

ELSPETH OPPERMANN*

Division of Geography, Lancaster Environment Centre, Lancaster University, Lancaster, LA1 4YW, UK

Discourse describes our reality, at once depicting and delimiting our understanding of the world and our place in it. Discourse can also be de-scribed; the script unpicked to reveal the contingency of what it includes and excludes. The discourse of adaptation to climate change is one such description. This article uses critical discourse analysis to describe and de-scribe the discourse of adaptation as (re)produced through one UK Climate Impacts Programme report. It identifies within the report a core, techno-scientific problematization of adaptation and a series of supplementary moments that are in ontological and epistemological tension with the core problematization, particularly regarding the conceptualization of knowledge and uncertainty, actors and agency, and the practices and politics that are understood to constitute adaptation. This tension is used to de-scribe the discourse; the supplementary moments highlight the limits of the core problematization, revealing points of extension or departure for the discourse. This is particularly apparent when these moments are articulated together into a supplementary, socio-systemic problematization. This demonstrates significant conceptual resonances with systems and complexity-based accounts of adaptation, including the discourse of sustainable adaptation, all of which offer further discursive resources for the extension or replacement of the discourse as produced in the report.

Keywords: adaptation; complexity; critical discourse analysis; discourse; risk; sustainable adaptation; systems; UK Climate Impacts Programme

1. Introduction

The last decade has seen a profusion of policy and advice on adaptation to climate change. As a Foucaultian account of discourse reminds us, the connections between language, knowledge and power mean that such policies are productive at far more than the technical level. They contribute to a conceptual description, or 'problematization', that at once depicts and delimits our understanding of reality and our subjectivity, profoundly shaping what practices and politics are understood to be possible, necessary or acceptable. As such, the problematization of adaptation shapes policy, yet policy also shapes the problematization of adaptation, depicting not only how adaptation is 'done', but also delimiting the possibility of doing it differently.

If the material practice of adaptation matters, this production of policies on adaptation to climate change and their increasing embeddedness in institutions and wider society calls for critique at the discursive level (Head, 2010). However, the discourse of adaptation as it occurs across the globe is too multiple and complex to consider here. Instead, this article explores a single case: a key report from within the discourse of the UK Climate Impacts Programme (UKCIP). In doing so, this article aims to demonstrate in detail how the problematization of adaptation is constituted in the report, and how this is significant for describing, as

■ *E-mail: e.oppermann@lancaster.ac.uk

ISSN: 1756-5529 (print)

depicting and delimiting adaptation practices. The report in question is the UKCIP technical report *Climate Adaptation: Risk, Uncertainty and Decision-making* (Willows and Connell, 2003).

UKCIP's discourse is significant because of the wider influence it has had on the discourse of climate change adaptation in the United Kingdom. It has worked directly with stakeholders from small to large companies, professional organizations, local authorities and regional climate change partnerships across multiple UK regions and sectors. A number of UK-based consultancies also use its literature (Connell, 2009; Willows, 2009). UKCIP's discourse has also influenced the formation of adaptation governance at national and international levels, including that of the European Environment Agency and the Intergovernmental Panel on Climate Change (ESYS Consulting, 2004; Colls et al., 2005; Turnpenny et al., 2005). Although UKCIP's discourse has developed substantially since the programme was created in 1997, the *Climate Adaptation* report has, from 2003 up to the present, laid the foundations of its approach through establishing a particular problematization of adaptation.[1]

The report appears to problematize adaptation through a techno-scientific epistemology centred on risk. This is articulated with a modernist ontology of actors and agency as separate and discrete, centralized in space and time. However, this core problematization is supplemented by moments in the text that imply a socio-systemic problematization. This introduces an ontology and epistemology that exist in tension with the report's dominant discourse by acknowledging the existence of networked actors and emergent social structures. It also refers to practices and politics that assume and build on these foundations. However, these are linguistically resolved back into the narrative of the report, such that they do not appear to disrupt the core problematization and hence the coherence of the discourse.

Nevertheless, these supplementary moments do introduce an alternative conceptual basis for accounting for adaptation. This article uses this tension to de-scribe the discourse in the report by revealing its contingency, thus opening it up to critique. A particular means of critique arises from the supplementary moments themselves. If drawn together, they formulate a supplementary problematization that not only indicates precise points where the core problematization might be extended, but in providing an alternative also provokes its replacement on these terms. Importantly for this special issue of *Climate and Development*, this supplementary problematization resonates with the emergent discourse of sustainable adaptation, making the latter a resource for discursive change within UKCIP if the supplementary problematization were to be pursued.

2. Adaptation, discourse and UKCIP

This research follows on from constructivist and post-structuralist approaches that are broadly concerned with the construction and contingency of discourses in the fields of environmental policy (Feindt and Oels, 2005), climate change science (Shackley and Wynne, 1995; Demeritt, 2001; Oels, 2005; Demeritt, 2006) and adaptation to climate change (Schipper, 2006; O'Brien et al., 2007). Although not always rendered specifically in terms of 'discourse', the common concern is the way in which narrative, concepts, language and policy fundamentally shape and limit how climate change and adaptation are problematized, and what policies result.

The profound impact that a given problematization has on policy outcomes is demonstrated by O'Brien et al. (2007) with regard to vulnerability. They identify two distinct conceptualizations of vulnerability that are 'outcome' and 'context' based. In the former, the consideration of particular kinds of knowledge and practice is limited to a 'scientific framing' of climate change. This gives rise to a narrow and linear understanding of causality and response, and to discretely measurable, materially focused policies (O'Brien et al., 2007). By contrast, the 'context'-based account of vulnerability is sensitive to the connections between and within society and environment, and uses this very connectivity as its mode of response.

O'Brien et al. (2007, p. 76) identify 'context' as that which is 'silenced' by the problematization of vulnerability as 'outcome'. Given the many links between adaptation and vulnerability discourses (Eriksen and O'Brien, 2007; O'Brien et al., 2007), this provokes the question: what does the problematization of adaptation include and exclude? Proponents of sustainable adaptation, including the authors mentioned above, do in fact critique adaptation discourse for often being 'techno-scientific' or 'outcome' based and tending to exclude the social (Eriksen and O'Brien, 2007; O'Brien et al., 2007). To redress this, they posit an account of adaptation as 'sustainable', which reproblematizes adaptation as the capacity of social systems to respond to the effects of climate change through being sensitive to changes in the social, political, economic and environmental 'context' (O'Brien and Leichenko, 2007, p. 31).

The divide in adaptation discourse between outcome/context and techno-scientific/socio-contextual may have some direct implications for UKCIP. Not least because UKCIP was explicitly tasked with bridging a perceived scientific/social divide in terms of who had expertise on adaptation and control of adaptation research agendas. Importantly, the claim by most of the literature on UKCIP is that it is remarkably attuned to the social, and celebrates its attempts to 'bridge the gap' between the contextual needs of those who enact adaptation and the techno-scientific knowledge of climate science (Hedger et al., 2006). More fundamentally, Hulme and Turnpenny (2004) see UKCIP as a good response to a post-normal science problem.[2] Lorenzoni et al. (2007) extend this by characterizing UKCIP as a boundary organization.[3] All of these accounts characterize UKCIP as an 'honest broker' (Lorenzoni et al., 2007, p. 73) in enabling a more open and cooperative engagement between scientific bodies and wider society to enable the production of knowledge useful to policy (Hulme and Turnpenny, 2004; Hedger et al., 2006; Lorenzoni et al., 2007).

However, 'bridging the gap' through procedures of engagement does not necessarily mean that the problematization of adaptation changes. Moreover, if it does, the unanswered question is: how? Lorenzoni et al. (2007, p. 73) do note that boundary organizations produce 'boundary objects' which help order both social and scientific worlds, but this aspect is not investigated further. This matters, because the accolade of 'honest broker' ignores the fact that all the knowledge and advice that UKCIP produces on this 'border' necessarily emerges from acts of interpretation or even creation, given how new the topic of adaptation to climate change is. In fact, the report considered here was mandated precisely to fill a void in UK government discourse on climate change adaptation (Connell, 2009; Willows, 2009). As such, the problematization of adaptation, itself a 'boundary object', needs to be considered. The first step to this analysis lieu in examining the key report *Climate Adaptation: Risk, Uncertainty and Decision-making* (Willows and Connell, 2003).

3. Theoretical and methodological premises

This research is itself situated within a broadly post-structuralist epistemology, which understands knowledge to be produced through discourse, and hence contingent (Campbell, 1998; Foucault, 1998; Howarth, 2000; Laclau and Mouffe, 2001). In this context, what is significant is the particular way a topic is rendered, as this constitutes knowledge (Foucault, 2002, 2003). 'Problematization' is Foucault's means of accounting for the conceptual logic that does this. It describes how the ontological and epistemological characterization of something as a 'problem' establishes and justifies proper objects and subjects for its 'solution', thereby constituting an understanding of what practices and political rationales are considered as legitimate in order to resolve the problem (Campbell, 1998; Foucault, 1998, 2002, 2003). Fairclough (2001, p. 128) gives the example of 'medicine' as a topic, but 'techno-scientific medical discourse' as problematizing medicine in a particular way, resulting in particular regimes of practices.

The differentiation between problematization and discourse here should be well noted, because it allows for the consideration that a problematization can change within the same discourse. Discourse in this sense is defined by Foucault as 'regularity in dispersion' which is explained by Laclau and Mouffe as expressing a degree of continuity in what the object of the discourse is, or as a relative continuity in the system of meaning of that discourse whilst the objects themselves might change (Campbell, 1998; Laclau and Mouffe, 2001, pp. 105–106; Hansen, 2006). Crucially, this leaves a space for change within a single discourse, such as the discourse of adaptation (re)produced by UKCIP.

Accounting for the problematization and its possibility to change was for Foucault very much tied to power relations, including power-knowledge. This puts a heavy emphasis on how a discourse is produced, or how it is constituted through the 'strategic codification' of power relations (Foucault, 1998, p. 96). Given the myriad of power relations that constitute UKCIP's discourse as a whole, and that the report is both produced by, and produces, power relations, this article recognizes a broader investigation than can be described here is required. However, given the strong positioning of the report as the official basis of UKCIP's discursive practices, it is an essential 'object' to consider in unpicking how UKCIP problematizes adaptation. The report can thus be used as an initial account of this problematization and as a comparative basis from which to analyse how this discourse has played out in practice, for example, in interaction with stakeholders, and in relation to the multiple boundary products produced as a result, such as adaptation strategies.

Before the report's discourse can be analysed for its practical and political implications, understanding the problematization at its core is necessary. Laclau and Mouffe's account of articulation is useful here. Articulation is understood as 'any practice establishing a relation among elements such that their identity is modified as a result of the articulatory practice'. As a result of being articulated, these elements are understood as 'moments' which together form a 'structured totality' that describes a discourse (Laclau and Mouffe, 2001, p. 105).

Although Foucault and Laclau and Mouffe tend to look at discourse as a 'strategic-codification' or 'articulation' of systems of meaning on a larger, generally society-wide scale, such articulation can also be achieved at the level of a text. Here, narrative and other modes of linguistic structuring affect the articulation of concepts or elements as moments, constituting a 'logic of explanation' that establishes a particular problematization (Campbell, 1998, p. 4; Howarth, 2000).[4] This allows for the report to be considered as a particular rendition of UKCIP's discourse, which is operationalized through the analysis of such linguistic techniques.

The process of problematization as a strategic articulation reflects the contingency of knowledge. Taken to its full extent, this presupposes an ontology of radical undecidability and the irreducibility of different accounts of the real (Howarth, 2000, p. 135). Thus, articulation is essential in that it not only includes but also excludes concepts in order to generate a coherent problematization. Such a position draws on Foucault's concern with the 'distribution' of 'discursive elements' 'with the things said and those concealed, the enunciations required and those forbidden' (Foucault, 1998, p. 100). Derrida demonstrated how this could be seen at the level of the individual text, which can be critically read to demonstrate its necessary 'limits' or 'points of closure' that establish coherence in the context of radical undecidability (Howarth, 2000, p. 135).

Discourse analysis offers useful tools to operationalize these theoretical foundations. Its tools of analysis, which draw from semantics, rhetoric and narrative theory, offer a means of traction for considering discourse at the level of the text (Jäger, 2001; Wetherell et al., 2001; Renkema, 2004; Johnstone, 2008). Such tools enable us to consider the precise way in which concepts are articulated such that they can form a logic through inclusions and points of closure and exclusion (Dijk, 2001). While CDA relies on a range of linguistic techniques, these are bound to the actual utilization of language in the report itself. These are made explicit through the discussion of findings below.

However, as the claim about the irreducibility of knowledge implies, discourses are far from pure, and far from stable. As we shall see, 'supplementary' concepts may be included in a text and a discourse, which are out of step with its core problematization. These are termed 'supplementary moments',[5] and draws on Derrida's two-fold account of 'the supplement'. First, as an additional term or identity that adds to, and brings fullness to, an existing term or identity; second, as necessarily also revealing the limitations and contingency of the existing term (Torfing, 1999; Hansen, 2006). The supplementary moment then is intended to examine how supplementary concepts might be linguistically articulated to a problematization that they are conceptually at odds with. Here, the manner of this articulation obscures these inconsistencies, and in doing so appears to give fullness to the problematization. In fact, they are doing the opposite: because of being articulated to the original problematization, these supplementary moments are in fact limit points; they are articulated in such a way that their conceptual differences remain indistinct and inert, such that they do not change the core problematization. However, if this conceptual disconnect is made clear, and the manner of their articulation to the core problematization unpicked, such moments can actually highlight the contingency of the core problematization, and even provide points of departure immanent to the text for its genuine extension or change.

On the basis of these theoretical and methodological foundations, the problematization the report produces will first be characterized by looking at its narrative as an 'articulation' of certain concepts as 'moments' (Laclau and Mouffe, 2001) within a conceptual logic (Campbell, 1998; Laclau and Mouffe, 2001). The problematization thus produced is then explored in more detail: first, the epistemology of adaptation will be considered; that is, how are knowledge and knowing understood in this discourse? Second, its ontology with, that is, who or what is assumed to exist and act in a way that is causing either the problem or the solution of adaptation? Third, the practices and politics this problematization of adaptation; that is, what kind of political rationale orchestrates particular practices, and what rationale do such practices constitute? In each of these sections, supplementary moments are briefly considered. The final section draws these together into a supplementary problematization, and uses this to discuss the implications for the validity and stability of the problematization of adaptation in the report. This de-scribes the discourse. It then suggests how these findings might be developed in terms of research into UKCIP's wider discourse, and how they be useful to any deliberate widening of its discourse.

4. The problematization of adaptation

The 'foreword' and 'context' sections of the report provide the first substantial narrative accounts of adaptation. In doing so, they play the crucial role of establishing the core problematization of adaptation based on the characterization of climate change as risk. Both sections are expressed in non-technical language and are rhetorically authoritative and persuasive. This creates an apparently logical and sufficient story of what adaptation is, leaving no explicit space for a critical reading. This initial 'shaping' narrative (Johnstone, 2008) makes the more detailed account of adaptation that follows easier to accept for the reader, because it appeals to this already established logic in order to appear comprehensive and concrete.

The problematization created by this narrative can be seen, for example, in the 'context' section. This begins with the claim that there is certainty that climate change is happening, but 'the exact extent and nature of changes in our climate remains uncertain' (Willows and Connell, 2003, p. 3). This is followed by the statement: 'Climate therefore represents a changing source of risk. Climate adaptation is about recognizing these altered risks, and taking decisions…'. Running this presupposition immediately after the discussion of the uncertainty of climate change, together with the use of 'therefore', implies that climate uncertainty is fully accounted for by the trope of 'risk'. Risk thus appears as the natural and sufficient way to understand the uncertainty of climate change.

The sentence continues: 'Climate adaptation is about recognizing these altered risks, and taking decisions that allow the likely impacts to be reduced or managed, and the opportunities exploited'. The authoritative use of the word 'recognizing', together with the first 'and', link risk as the account of the problem apparently straightforwardly with the account of its solution. Interestingly, it is not specified here how the term 'risk' is being used. The simplicity of the move from problem to solution implies the risks are knowable, and an account of risk as precise or 'known' probabilities. It is noted that the risks are changing, although these are only 'altered', implying minor change and hence accessible knowledge of climate change as risk.

Thus, in this short sentence, a logic is established that calibrates the uncertainty of climate change with risk in a way that implies that risk knows uncertainty in full, and that this knowledge is a sufficient and straightforward basis for decision-making that will enable successful adaptation. This problematization sets up the core conceptual logic of adaptation: an epistemology of climate change impacts based on risk, and an ontology that assumes that these risks can not only be known but also direct subsequent actions, entailing practices and politics built on trying to identify risk as the primary basis of making decisions.

However, in the main body of the report, there are significant moments of more open, tentative and limited claims about the use of risk-based knowledge as a basis for decision-making and solutions to climate change uncertainty. These are sometimes coupled with the suggestion of alternative ways of knowing, and alternative accounts of and the proper object means of knowledge for the purpose of adaptation. However, when these moments appear to compromise the risk-based decision-making problematization, they are resolved through a variety of linguistic techniques back into the narrative, thus securing the apparent integrity of the problematization.

5. Epistemology

At the core of any problematization are some basic assumptions about what knowledge is and how it can be attained. In narrative and rhetorical terms, as well as in terms of the sheer volume of risk-based technical information, it is clear that in this report knowing is framed predominantly by science and particularly risk. Risk and risk technologies become both the means of comprehending climate change impacts and a large part of knowing what to do to adapt.

Indeed, risk becomes the prism through which both knowledge and uncertainty are understood and valued; both the means of knowing and the end of knowledge are in risk. This occurs when, for example, the report distinguishes risk from the kind of epistemic uncertainty most associated with climate change (Willows and Connell, 2003, p. 43). Uncertainty is stated to be 'the result of a lack of knowledge of either the probability of an event, or its consequences' (p. 50), whereas 'good knowledge of both' enables accurate characterization of risk. From these crucial definitions, risk is set up as not only distinct from uncertainty, but almost as its opposite. The report states this overtly when it says 'Uncertainty describes the quality of our knowledge concerning risk' (p. 43). This negative account of uncertainty as the absence of risk-based knowledge renders the latter as the epistemic ideal, making risk appear a sufficient basis for 'knowing' adaptation.

Although the report sometimes signals concern that epistemic uncertainty sits beyond the capacity of risk-based techniques, uncertainty is subsequently brought back into the epistemology of risk through the related definition of 'risk assessment'. Risk assessment is defined as involving

> either quantitative or qualitative techniques and information to describe the nature of the probability component of the risk.... Qualitative techniques are particularly useful in circumstances where we lack knowledge of the probabilities. (p. 44)

Here, qualitative knowledge is used to create proxy risk-based knowledge. The core epistemology thus remains in place, as other epistemologies act as subsidiary means of providing information while the 'knowing' is actually done through risk, rendering their own epistemologies and related ontologies invisible.

The glossary of the report supports this prioritization of risk, and gives a definition of adaptation itself directly in these terms as 'The process or outcome of a process that leads to a reduction in harm or risk of harm, or realization of benefits associated with climate variability and climate change....' (p. 111) From this we see that risk, and the harm/benefit metric, places the focus on knowing the 'outcome', or 'product', not just of climate change impacts but of adaptation itself.

5.1. Supplementary epistemology

Despite the narrative that runs through the text articulating risk and risk-based decision-making with knowing and acting, there seems to be a tension within the report as to the degree to which knowledge as certainty can be accessed through risk. The report contains some striking moments that supplement the techno-scientific framing. The first is the consideration that even though climate change is knowable as risk, there are elements that exceed risk and are really unknowable (p. 50). This undermines the validity of the problematization of climate change-as-uncertainty and uncertainty-as-risk, and thus of risk technologies as being the appropriate and sufficient means by which to adapt.

Similarly, although determining the risk of an event as an outcome remains the key objective in order for decisions to be made, the basis for a decision is occasionally expressed as knowing the characteristics of the system that is at risk and the processes with which it is engaged which may render it vulnerable. While this is not incompatible with risk in principle, it does shift the focus of knowing from outcome to context, and from the externally generated 'event' to internally generated generic vulnerability. In the text, these moments of an alternative object of knowing are linked immediately back to the concept of risk through the designation of these as 'risk control points' (p. 47). As such, any knowledge that is gained about the system is brought back to the problematization's epistemology: valid as knowledge only in so far as it can be used to produce a risk-based outcome. The result is the occlusion of the internal system as a focus of knowledge in itself, and also as the means of knowing the threat of climate change and adaptation.

6. Ontology

The supplementary moments that disrupt the outcome-based account of knowing raise the question of what should be known, by whom and to what purpose. The report does account for climate change as an object of knowledge, but for the purpose of adaptation understands the nexus between climate change and exposure to it as the object of knowledge. Crucially, the report implicitly and explicitly understands the subjects of risk as organizational entities such as government agencies, private sector companies and third sector organizations. These are, perhaps unsurprisingly, accounted for in generally liberal terms, explicitly described as discrete organizations with individualized aims and objectives (p. 11).

The link between epistemology and ontology within a problematization comes into play here.

An ontology of discrete subjects is entailed by the epistemology of risk. First, in its desire to identify future outcomes, risk both assumes and requires a discrete conceptualization of social space; it requires particular actors who can not only 'do' the risk assessment, but also make an event's likelihood meaningful as risk by tying it as an impact to a particular, and therefore discrete entity. Second, the desire to assess risk also requires that this entity projects itself forward in time, thus assuming the stability of the subject across time, as a pre-emptive assessment of a future event is being given meaning by the current state. Moreover, the responses to the risk are assumed to act on these presumed principles, and thus ensure the continuation of an essentialized self. The problematization of adaptation as risk thus produces an ontology of the discrete and reified subject.

The articulation of such a subject with organizational entities creates a particular ontology of agency. While these entities might be assumed to be the subjects of risk, the subjects explicitly given agency in the report are described as the 'decision-makers' within these entities. Who is a decision-maker is a little vague, but is associated in the text with policy-makers and top management (p. 11). The term assumes that the making of decisions is an executive role, excluding broader understandings of agency as distributed, and thus a broader understanding of the generation of adaptation decisions.

Indeed, this centralization of agency within the organization is reinforced through an explicit exclusion of the social. Society is seen as an object, rather than as subject with the ability to think or act. Early on in the text, society is characterized as something the decision-maker can have knowledge 'of' (p. 3), thus detaching the decision-maker from society, and excluding society from having agency. Much later a very different claim about the 'knowability' of society is made, yet it also explicitly establishes the same exclusion. The statement made is that both society and the economy are fundamentally unknowable in the sense of not being predictable (p. 50), a claim that is valid only within the epistemology of outcome-based knowledge. This excludes 'society' not only as object but also as subject from the problematization of adaptation.

The account of the subject and its agency as limited and discrete, articulated together with the assumption that climate change and adaptation are calculable, also seem to imply the possibility of, and desire for, the 'best decision' as a pre-determined outcome. Adaptation is in fact explicitly defined in the glossary in terms of positive outcomes, as it is seen as either the process that leads to, or the outcome of a process that equates to, the reduction of risk or harm and the production of benefits. While an account of process is included, it is rendered subsidiary to outcome in that it is defined in terms of results as harm or benefit. This again places the outcome as the objective, and hence as the object of knowledge.

On this basis, it is perhaps not surprising that there is no substantive definition of 'adaptation' per se in the glossary beyond its coupling with 'climate', as knowledge of adaptation is also focused on the external, future event, rather than on any substantive process. This can only make sense if there is assumed to be a means of response appropriate for any event, to achieve any adaptation decision. Unsurprisingly perhaps, this gap tends to be implicitly filled by liberal norms of organizational practice, which the report articulates to adaptation indirectly through its ontology of discrete units controlled by decisions centralized in space and in time.

6.1. Supplementary ontology

However, there are several principles mentioned that nuance this ontology and in doing so act as supplementary moments. These offer an account of what substantively constitutes 'good' adaptation, yet are conceptually more akin to the supplementary systems-based epistemology discussed above. First, although the report explicitly considers that organizations are ultimately responsible to themselves, they are nonetheless imagined as located in networks with shared interests or roles, such as stakeholder groups and 'sectors'. As such, they are asked to involve these in their decision-making process (p. 66). Linked to this ontology is a desire to avoid 'knock-on effects' on others that limits their options of adaptation (p. 17). These principles appear to recognize and counteract the limitations of decisions centralized in space by assuming a more complex account of society as made up of relations. However, these points are not developed in terms of more precise recommendations or practices.

Second, although the report's core problematization requires decisions be centralized in time in order to pre-determine how to adapt to an impact, it also specifies that decisions should be iterative, and that management should adopt practices of 'adaptive-management' (pp. 28, 71). The report clearly notes that the

> emphasis of this framework on adaptive management strategy supported by post-decision monitoring and appraisal is essentially a defence against uncertainty, recognizing that for many aspects of climate change adaptation, uncertainty will be significant. (p. 53)

Here the management strategy itself appears as a core account of adaptation, rather than the specific decisions themselves or the risk-based framework through which they are made. Yet this process, and the mechanisms and the qualities of relations it requires (Forsyth, 2010), seems to warrant far more focus and detail than the report provides. Instead, management, organizational and social aspects are left largely un-described, and his silence excludes them from the conceptualization of adaptation.

The supplementary ontology of epistemic uncertainty is given further weight when the report establishes the principle that decisions should ensure the flexibility of future decisions; 'no regret' and 'low regret' options 'should' be identified 'at the outset' (p. 30). The concept of adaptive management is also directly articulated with this, as it 'should be directed towards an overall strategic objective. In all cases an objective must be to keep open possible future options, that is, avoid decisions that constrain future options for adaptation' (p. 30). This is more than a decision being appropriate for a variety of foreseen outcomes; it intimates that there are unforeseen outcomes and therefore that the more capacity a decision retains, however latent, the better it is. This would seem to link better to a system-based ontology of contingent emergence than a scientific ontology of outcome-seeking certitude and objectivism.

In sum, these supplementary moments of stakeholder engagement, adaptive management and open decisions, suggest that the key characteristics of successful adaptation are flexibility, being effective and complimenting adaptation in other sectors. All imply a more contextual, systems- and capacity-based ontology of the problem and of the solution, particularly in the understanding of the subject or actor that responds. However, while these moments appear to nuance the argument within the text, they are not explored in any practical detail, and are either tentatively or ambiguously worded, or not directly connected to the core narrative and ontology of who acts, how and on what, to secure adaptation. As a result, they have little effect on the core problematization of adaptation.

7. Practices and politics

The core ontology and epistemology identified above describe the boundaries of the proper objects of concern and ways of knowing them, and prescribe which actors participate in adaptation and what kind of practices are idealized and allowed. In doing so, they institute a logic which acts as a political rationale, giving meaning and the appearance of necessity to the practices that are called for (Campbell, 1998; Laclau and Mouffe, 2001).

One core element of this political rationale is the characterization of climate change as a threat of at least as a problem. This is established through the frequent connections in the report to the UKCIP climate scenarios, and explications of the kinds of impacts climate change will have. It produces a 'need' to adapt which demands a solution, which is provided by the risk-based framework. In this sense, the risk-framework is itself the foundational element of the practice of adaptation. This practice is assumed by the report to be primarily tied to the role of the decision-maker. As such, the practice of adaptation lies in the act of 'decision-making' based on a risk calculus. To revert to one of the opening claims, '[climate] adaptation is about recognizing these altered risks, and taking decisions that allow the likely impacts to be reduced or managed, and the opportunities exploited' (p. 3). Importantly, a temporal framing is also established here which assumes, and therefore limits, adaptation practices to considering future, 'likely impacts' and excludes real tune, actual impacts. Together with the use of the term 'risk', this casts the practice of adaptation into a predictive rather than responsive practice, and makes understanding the external, rather than the internal, its focus.

While the report's supplementary ontology may hint at what it considers a 'correct' attitude to risk, namely the principles of mutual consideration across time and space, there are other principles that enable these to be sidelined. The document is very careful to allow the primary concern to be the 'objectives of the decision-maker' (p. 53), reinforcing practice and politics of a liberal, laissez-faire value system. Likewise, it also recognizes and accepts institutional 'constraints', characterizing them positively, if perhaps rather euphemistically, as providing 'focus' to decision-making (p. 14). This not only allows but legitimizes a spatially and temporally limited political rationale and practices.

7.1. Supplementary practices and politics

Although not woven strongly into the text, there are moments that indicate a supplementary account of practices and politics that arises from an appreciation of social systems. There is, for example, a concern to avoid knock-on effects by considering current and future stakeholders, and keeping options open. Interestingly, these principles are noted as 'sustainability' in the text (p. 17). This indicates a politics that considers society or networks to be at least as valuable as the individual organization, as well as a practical understanding that the capacities of the organization are dependent on its relations. This implies a shift in the problematization of adaptation from the decisions that are made to the capacity and process of making decisions, including the capacity of the system to function as such. The presence of these supplementary moments indicates that 'sustainable' social systems may well be a productive way to reproblematize adaptation and in doing so poses fundamental questions about the sufficiency of the extant problematization.

It is clear there is a considerable amount of conceptual tension between this techno-scientific, predictive or outcome-based rationale, and a supplementary rationale based on irreducible uncertainty, provoking practices of adaptive management and other modes of response that maintain flexibility. However, these concepts are retained within the core problematization. For example, adaptive management is described as simply revisiting decisions at regular intervals, leaving the objective of defining and predicting for the purpose of planning a response. Now that these supplementary ideas have been isolated from their linguistic placement, however, the discussion below will consider their conceptual implications.

8. Discussion

All the key aspects of the problematization of adaptation in the report are marked by the presence of supplementary moments. However, although included in the text, these are rendered inert by the manner in which they are linguistically articulated to the report's narrative. These moments do, nonetheless, introduce supplementary conceptualizations of epistemology, ontology, practices and politics. These appear to share several characteristics, and could be taken to form a supplementary problematization of adaptation that extends yet also exceeds the report's core problematization.

To begin with, the supplementary epistemology is concerned with knowing the social, particularly in terms of networks or systems. In doing so, the supplementary epistemology suggests a problematization of adaptation based on questions of complexity, contingency and emergence. Because of the application of such concepts beyond the biophysical to the social, they exceed risk in as much as the very values by which risk is assessed are themselves understood as emergent, undermining the validity of outcome-based assessments. In this context, the epistemology might be associated with an ontology of irreducible uncertainty and the foundational assumption of ignorance that the system in its current, let alone future, state cannot be predicted, particularly from a vantage point centralized in space and time (Cilliers, 1998; Luhmann, 1998).

This changes the way knowledge is understood and what is problematized as a result. If assessments of impacts on a system could not assume a stable 'internal' system in seeking to predict an 'external' threat, what is problematized is the system's ability to know itself, its own connections and limits. Moreover, prediction would be replaced with the ability to know the event as it occurs and unfolds. These means of knowing are excluded by the report's outcome-oriented epistemology.

This different epistemology requires the establishment and maintenance of a capacity to sense, understand and communicate knowledge quickly throughout a system. This entails in ontology that substantively accounts for the nature of the system and its ties, including social and organizational relations and structures, and their contingency (Cilliers, 1998).

Within this ontology, the subject also changes. 'Individuals' such as decision-makers and their organizations are recast as nodes within a system. Given the node can only act by virtue of its relations or capacities within the system as a whole, the 'subject' of adaptation tends towards the system itself. The question of agency and identity is then addressed to the node-as-part of the system, or the system as made up of nodes and their relations. There is an enormous gap between this account of distributed agency and the ontology of 'the' decision-maker, who problematizes adaptation as centralized planning, which assumes centralized identity and agency.

Reproblematizing adaptation in terms of the system as an emergent object and subject entails a political rationale to secure the system as such. This rationale might entail the supplementary politics discussed earlier, of mutual consideration and trust across space and time, resulting in practices designed to enable this, such as communication and cooperation. Given the complexity of the system, this trust might also be seen as essential to enabling actual 'decisions' to be decentralized across space and time in order to ensure the appropriateness and timeliness of the response.

This account reinterprets and undermines the validity of the report's core political rationale. For example, the value of risk-assessment practices changes from the outputs of the assessment process to the process of making the assessment itself because of the ties of communication and system-awareness that assessment requires. Moreover, in the context of social complexity, encouraging or simply allowing narrow organizational interest, or even sectoral or national interest, would be seen as debilitating to the wider system, and hence general capacity. Indeed, because it might create conflicting areas of order, thus increasing overall disorder (Urry, 2005), it may be seen as a threat rather than a solution to successful adaptation on a society wide account.

These supplementary moments are significant sources of critique where they provide substantive indications of how the account of adaptation might be extended. By removing these from their linguistic articulations to the narrative and hence core problematization in the text, it is possible to identify resonances with alternative discourses of adaptation that are similarly sensitive to the social, including 'sustainable adaptation' (see, e.g. Pelling et al., 2008; Brockhaus and Kambire, 2009; Inderberg and Eikeland, 2009). These take the nature of the social as central to enabling adaptation. Indeed, Eriksen and O'Brien go so far as to say that 'adaptation is a social development issue as much as (if not more than) an environmental and technological issue' (O'Brien et al., 2007, pp. 348–349).

The understanding of the social as a complex system also links to other discourses, such as systems and complexity theory, complex adaptive systems thinking and organizational learning, which share its general ontology and epistemology. These all offer resources to extend and deepen such a problematization of adaptation, which is so minimally accounted for by the techno-scientific problematization. For example, adaptation problematized as systemic capacity would explicitly theorize questions of integrity and flexibility, particularly as a product of increased levels of communicability and willingness to change (Folke et al., 2002; Adger et al., 2009; Pelling et al., 2008).[6]

The presence of these supplementary moments in the report points to the need to investigate what effect they might already have had across UKCIP's entire discourse. Key questions in this respect are the following: what happens to these supplementary moments outside of the controlling effects of the linguistic structuring of the report in the text? Do they become more pronounced when the report's discourse is reproduced in more flexible verbal and interactive modes as UKCIP interacts with stakeholders? Are these supplementary moments picked up by particular subjects and used to make links to other discourses, driving the extension or even replacement of the original problematization? Do practices of adaptation highlight these conceptual tensions, and what effect does this have on UKCIP's discourse? If there is no change, why not? Might these supplementary conceptualizations be dropped by subjects as 'too difficult' to reconcile with the rationale established by the report, or with a wider rationale? What is the conceptual relationship between UKCIP's discourse and deeply entrenched discourses of liberal government and the value of techno-scientific knowledge?

The investigation of such questions would need to include further UKCIP literature and practices. In particular, the following areas seem likely to enhance the visibility of a problematization of adaptation in socio-systemic terms: UKCIP's engagements with stakeholders in general, and partnerships in particular, such as the regional climate change partnerships; the development of more process-based approaches, such as National Indicator-188; and the probabilistic climate projections of UKCP'09, to see whether its increased accounting for uncertainty drives the production of adaptation practices more oriented towards complexity.

How these areas of discursive practice reflect, deepen or resolve the conceptual tensions identified in the report will have fundamental effects on the problematization of adaptation and the practices that result, not just within UKCIP, but across the UK. As Foucault notes, the presence of a supplementary problematization within a discourse has the potential to generate at best 'contrary objectives' (Foucault, 1998, p. 100), and at worst an immanent, if not imminent, 'crisis of governmentality' (Foucault, 2008, p. 68). Such crisis is if ignored, destructive but may also be productive if its conceptual underpinnings are observed and explored, allowing a rearticulation of the problematization to face these issues.

9. Conclusions

This article seeks to describe UKCIP's discourse of adaptation as it appears in the core UKCIP technical report, *Climate Adaptation: Risk, Uncertainty and Decision-making* (Willows and Connell, 2003). It identifies the report as having a core techno-scientific problematization of adaptation based on risk. This is linked to an ontology that occludes the social as the object and subject of adaptation, particularly by centralizing social time and social space through its focus on predictive knowledge, 'the decision' and 'the decision-maker'. The result is a politics of adaptation based on the individualized needs of economic or governmental entities, reflecting liberal norms.

However, this problematization is nuanced by several supplementary moments. These challenge the epistemology of risk as a means of knowing, particularly where this excludes consideration of the social system that responds. They extend the ontology to consider the contingency of social systems, entailing a broader account of the object and subject of adaptation, including actors and agency, as distributed throughout the system. These foundational aspects correspond to supplementary politics and practices based on systemic cooperation and communication. Further, these moments can be drawn together into a supplementary problematization of adaptation in terms of the social system's capacity to adapt, substantiated in terms of ensuring the integrity and flexibility of the system.

This article uses CDA to uncover these supplementary moments, but also to show that, despite their presence, they are not articulated into the core narrative of the report or are articulated in the manner that renders them subsidiary. As such, they remain in conceptual tension with its core problematization, which is particularly evident if they are themselves rearticulated into a supplementary problematization. In unpicking this tension, this article de-scribes the discourse presented by the report, showing its areas of inclusion and exclusion. This demonstrates that UKCIP's discourse of adaptation is open to question, and that the supplementary problematization offers points from which to extend and replace the current problematization. The productiveness of these points is enhanced by their resonances with alternative discourses of adaptation, including the discourse of 'sustainable adaptation,' and discourses of complexity and systems-based thinking.

Notes

1. The report also has a less technical online version called the *Adaptation Wizard*. In observing and discussing the more detailed problematization established in the technical report we may perhaps arrive at the clearest understanding of this problematization.
2. A post-normal science problem as defined by Funtowicz and Ravetz's *Science for the Postnormal Age* (*Futures*, 25, 739–755).
3. The term is based on the account of two papers by Guston (1999, 2001).
4. Here, Campbell draws on Hayden White (1987), 'The Content of Form: Narrative Discourse and Historical Representation.'
5. This term is used by the author in lieu of finding a more precise term or existing theorization of this linguistic operation at this point in time. The author is aware that there is a wide array of uses of Derrida's supplement which should be explored here, and would appreciate any comments or feedback.
6. There is no intention here to present this particular supplementary problematization as 'good' or to make a normative case for its inclusion in UKCIP's problematization of adaptation.

References

Adger, W. N., Dessai, S., Goulden, M., Hulme, M., Lorenzoni, I., Nelson, D., Otto-Naess, L., Wolf, J. and Wreford, A., 2009. Are there social limits to adaptation to climate change? *Climatic Change*, 93(3–4). 335–354.

Brockhaus, M. and Kambire, H., 2009. Decentralization: a window of opportunity for successful adaptation to climate change? *Adapting to Climate Change: Thresholds, Values, Governance*, N. Adger, I. Lorenzoni and K. O'Brien (eds). Cambridge University Press, Cambridge.

Campbell, D., 1998. *Writing Security: United States Foreign Policy and the Politics of Identity*. University of Minnesota Press, Minneapolis.

Cilliers, P., 1998. *Complexity and Postmodernism: Understanding Complex Systems*. Routledge, London.

Colls, A., Erhard, M., Fritsch, U., van Miuner, J. and Voigt, T., 2005. *Vulnerability and Adaptation to Climate Change in Europe*. EEA Technical Report, European Environment Agency, Copenhagen.

Connell, R., 2009. Interview with Richenda Connell, Oxford, 30 January 2009.

Demeritt, D., 2001. The construction of global warming and the politics of science. *Annals of the Association of American Geographers*, 91(2). 307–337.

Demeritt, D., 2006. Science studies, climate change and the prospects for constructivist critique. *Economy and Society*, 35(3). 453–479.

Dijk, T. V., 2001. Principles of critical discourse analysis. *Discourse Theory and Practice: A Reader*, S. T. M. Wetherell and S. J. Yates (eds). Sage, London. 300–317.

Eriksen, S. and O'Brien, K., 2007. Vulnerability, poverty and the need for sustainable adaptation measures. *Climate Policy*, 7. 73–88.

ESYS Consulting, 2004. *Review of UKCIP: Final Report*.

Fairclough, N., 2001. The discourse of new labour: critical discourse analysis. *Discourse as Data: A Guide for Analysis*, S. Yates, M. Wetherell and S. Taylor (eds). Sage and The Open University, London and Milton Keynes. 227–266.

Feindt, P. and Oels, A., 2005. Does discourse matter? Discourse analysis in environmental policy making. *Journal of Environmental Policy & Planning*, 7(3). 1–13.

Folke, C., Carpenter, S., Elmqvist, T., Gunderson, L., Holling, C. S. and Walker, B., 2002. Resilience and sustainable development: building adaptive capacity in a world of transformations. *Ambio*, 31(5). 437–440.

Forsyth, T., 2010. Panacea of paradox? Cross-sector partnerships, climate change, and development. *Wiley Interdisciplinary Reviews: Climate Change*, 1(5). 683–696.

Foucault, M., 1998. *The Will to Knowledge: The History of Sexuality*. Vol I. Penguin Books, London.

Foucault, M., 2002. *The Order of Things*. Routledge, London.

Foucault, M., 2003. *Society Must be Defended: Lectures at the College de France 1975–1976*. Picador, New York.

Foucault, M., 2008. *The Birth of Biopolitics: Lectures at the College de France 1978–1979*. Palgrave Macmillan, Basingstoke.

Funtowicz, S. O. and Ravetz, J. R., 1993. Science for the postnormal age. *Futures*, 25(7). 739–755.

Guston, D. H., 1999. Stabilizing the boundary between US politics and science: the role of the Office of Technology Transfer as a boundary organization. *Social Studies of Science*, 29(1). 87–111.

Guston, D. H., 2001. Boundary organizations in environmental policy and science: an introduction. *Science, Technology and Human Values*, 26(4). 399–408.

Hansen, L., 2006. *Security as Practice: Discourse Analysis and the Bosnian War*. Routledge, Abingdon.

Head, L., 2010. Cultural ecology: adaptation – retrofitting a concept? *Progress in Human Geography*, 34(2). 212–234

Hedger, M. M., Connell, R. and Bramwell, P., 2006. Bridging the gap: empowering decision-making for adaptation through the UK Climate Impacts Programme. *Climate Policy*, 6. 201–215.

Howarth, D., 2000. *Discourse*. Open University Press, Maidenhead.

Hulme, M. and Turnpenny, J., 2004. Understanding and managing climate change: the UK experience. *Geographical Journal*, 170(2). 105–115.

Inderberg, T. H. and Eikeland, P. O., 2009. Limits to adaptation: analysing institutional constraints. *Adapting to Climate Change: Thresholds, Values, Governance*, N. Adger, I. Lorenzoni and K. O'Brien (eds). Cambridge University Press, Cambridge.

Jäger, S., 2001. Discourse and knowledge: theoretical and methodological aspects of a critical discourse and dispositive analysis. *Methods of Critical Discourse Analysis*, R. Wodak and M. Meyer (eds). Sage, London. 32–62.

Johnstone, B., 2008. *Discourse Analysis*. Blackwell, Oxford.

Laclau, E. and Mouffe, C., 2001. *Hegemony and Socialist Strategy: Towards a Radical Democratic Politics*. Verso, London.

Lorenzoni, I., Jones, M., and Turnpenny, J., 2007. Climate change, human genetics and post-normality in the UK. *Futures*, 39. 65–82.

Luhmann, N. 1998. *Observations on Modernity*. Stanford University Press, Stanford.

O'Brien, K., Eriksen, S., Nygaard, L. and Schjolden, A., 2007. Why different interpretations of vulnerability matter in climate change discourses. *Climate Policy*, 7. 73–88.

O'Brien, K. and Leichenko, R., 2007. *Human Security, Vulnerability, and Sustainable Adaptation*. Human Development Report 2007/2008 Occasional Paper: Fighting Climate Change: Human Solidarity in a Divided World, HDROO paper.

Oels, A., 2005. Rendering climate change governable: from biopower to advanced liberal government? *Journal of Environmental Policy & Planning*, 7(3). 185–207.

Pelling, M., High, C., Dearing, J. and Smith, D., 2008. Shadow spaces for social learning: a relational understanding of adaptive capacity to climate change within organisations. *Environment and Planning A*, 40. 867–884.

Renkema, J., 2004. *Introduction to Discourse Studies*. John Benjamins, Amsterdam.

Schipper, L., 2006. A conceptual history of adaptation in the UNFCCC process. *Review of European Community and International Environmental Law*, 15(1). 82–92.

Shackley, S. and Wynne, B., 1995. Global climate change: the mutual construction of an emergent science-policy domain. *Science and Public Policy*, 22(4). 218–230.

Torfing, J., 1999. *New Theories of Discourse: Laclau, Mouffe, and Zizek*. Blackwell, Oxford.

Turnpenny, J., Haxeltine, A., Lorenzoni, I., O'Riordan, T. and Jones, M., 2005. *Mapping Actors Involved in Climate Change Policy Networks in the UK*. Tyndal Centre working paper no. 66.

Urry, J., 2005. The complexity turn. *Theory Culture Society*, 22(1). 1–15.

Wetherell, M., Taylor, S. and Yates, S. J. (eds). 2001. *Discourse Theory and Practice: A Reader*. Sage, London.

White, H., 1987. *The Content of the Form: Narrative Discourse and Historical Representation*. The John Hopkins University Press, Baltimore.

Willows, R., 2009. Interview with Robert Willows, Reading, 2 February 2009.

Willows, R. and Connell, R., 2003. *Climate Adaptation: Risk, Uncertainty and Decision-Making*. UKCIP Technical Report, UKCIP, Oxford.

For Product Safety Concerns and Information please contact our EU representative GPSR@taylorandfrancis.com Taylor & Francis Verlag GmbH, Kaufingerstraße 24, 80331 München, Germany

Batch number: 10732714

Printed by Printforce, the Netherlands